T0174679

Bayesian Reasoning and Gaussian Processes for Machine Learning Applications

Bayesian Reasoning and Gaussian Processes for Machine Learning Applications

Edited by
Hemachandran K
Shubham Tayal
Preetha Mary George
Parveen Singla
Utku Kose

CRC Press
Taylor & Francis Group
Boca Raton London New York

CRC Press is an imprint of the
Taylor & Francis Group, an **informa** business

A CHAPMAN & HALL BOOK

First edition published 2022
by CRC Press
6000 Broken Sound Parkway NW, Suite 300, Boca Raton, FL 33487-2742

and by CRC Press
4 Park Square, Milton Park, Abingdon, Oxon, OX14 4RN

CRC Press is an imprint of Taylor & Francis Group, LLC

© 2022 selection and editorial matter, Hemachandran K, Shubham Tayal, Preetha Mary George, Praveen Singla and Utku Kose; individual chapters, the contributors

Reasonable efforts have been made to publish reliable data and information, but the author and publisher cannot assume responsibility for the validity of all materials or the consequences of their use. The authors and publishers have attempted to trace the copyright holders of all material reproduced in this publication and apologize to copyright holders if permission to publish in this form has not been obtained. If any copyright material has not been acknowledged please write and let us know so we may rectify in any future reprint.

Except as permitted under U.S. Copyright Law, no part of this book may be reprinted, reproduced, transmitted, or utilized in any form by any electronic, mechanical, or other means, now known or hereafter invented, including photocopying, microfilming, and recording, or in any information storage or retrieval system, without written permission from the publishers.

For permission to photocopy or use material electronically from this work, access www.copyright.com or contact the Copyright Clearance Center, Inc. (CCC), 222 Rosewood Drive, Danvers, MA 01923, 978-750-8400. For works that are not available on CCC please contact mpkbookspermissions@tandf.co.uk

Trademark notice: Product or corporate names may be trademarks or registered trademarks and are used only for identification and explanation without intent to infringe.

Library of Congress Cataloging-in-Publication Data
Names: K., Hemachandran, editor. | Tayal, Shubham, editor. | George,
Preetha Mary, editor. | Singla, Parveen, editor. | Kose, Utku, 1985- editor.
Title: Bayesian reasoning and Gaussian processes for machine learning
applications / edited by Hemachandran K., Shubham Tayal, Preetha Mary
George, Parveen Singla, Utku Kose.
Description: First edition. | Boca Raton : Chapman & Hall/CRC Press, 2022. |
Includes bibliographical references and index. | Summary: "The book Bayesian Reasoning and Gaussian Processes for Machine Learning Applications talks about Bayesian Reasoning and Gaussian Processes in machine learning applications. Bayesian methods are applied in many areas such as game development, decision making and drug discovery. It is very effective for machine learning algorithms for handling missing data and for extracting information from small datasets. This book introduces a statistical background which is needed to understand continuous distributions and it gives an understanding on how learning can be viewed from a probabilistic framework. The chapters of the book progress into machine learning topics such as Belief Network, Bayesian Reinforcement Learning etc., which is followed by Gaussian Process Introduction, Classification, Regression, Covariance and Performance Analysis of GP with other models. This book is aimed primarily at graduates, researchers and professionals in the field of data science and machine learning"—Provided by publisher.
Identifiers: LCCN 2021052540 (print) | LCCN 2021052541 (ebook) |
ISBN 9780367758479 (hardback) | ISBN 9780367758493 (paperback) |
ISBN 9781003164265 (ebook)
Subjects: LCSH: Bayesian statistical decision theory—Data processing. |
Gaussian processes—Data processing. | Machine learning.
Classification: LCC QA279.5 .B43 2022 (print) | LCC QA279.5 (ebook) |
DDC 006.3/101519542—dc23/eng/20220128
LC record available at https://lccn.loc.gov/2021052540
LC ebook record available at https://lccn.loc.gov/2021052541

ISBN: 978-0-367-75847-9 (hbk)
ISBN: 978-0-367-75849-3 (pbk)
ISBN: 978-1-003-16426-5 (ebk)

DOI: 10.1201/9781003164265

Typeset in Palatino
by codeMantra

Contents

Preface

When we look into the past years, we can see an explosion in the applications of machine learning, particularly in e-commerce, social media, gaming, drug discovery, and many other verticals. These applications were focused on predictive accuracy and involved huge amounts of data. Bayesian methods give superpowers to machine learning algorithms, in handling missing data and in extracting information from small data sets. Bayesian methods help estimate uncertainty in predictions, which enhances the field of medicine. They allow to compress models a hundredfold and to automatically tune hyperparameters by saving time and money. In *Bayesian Reasoning and Gaussian Processes for Machine Learning Applications*, we discuss the basics of Bayesian methods, define probabilistic models, and make predictions using them. We discuss the automated workflow and some advanced techniques on how to speed up the process. We also look into the applications of Bayesian methods in deep learning and to generate images.

This book is designed to encourage researchers and students from multiple disciplines toward the arena of applications of machine learning. It aims to introduce a statistical background needed to understand continuous distributions and how learning can be viewed from a probabilistic framework. It also discusses machine learning topics such as belief network, Bayesian reinforcement learning, Gaussian process with classification, regression, covariance, and performance analysis of Gaussian processes with other models. This book is segmented into ten chapters.

Chapter 1 deals with the introduction of Naive Bayes – a collection of algorithms based on Bayes theorem – and its applications. It's a simple technique for constructing classifiers. Chapter 2 gives insights on different regression analyses in supervised learning. Chapter 3 throws light on different methods to predict the performance analysis of various machine learning applications. Chapter 4 discusses on belief networks and its applications. Chapter 5 describes reinforcement learning using Bayesian algorithms with applications. Chapter 6 intuits on alerting system for gas leakage in pipelines. Chapter 7 gives a perception on non-parametric models for biological networks. Chapter 8 provides a deep understanding on generating various types of graphical models via MARS. Chapter 9 is an acumen on financial applications of Gaussian processes and Bayesian optimization. Chapter 10 gives a panoramic view on Bayesian Network interface on diabetes risk prediction data. The book gives an insight into how new drugs can cure severe diseases with Bayesian methods.

We hope our attempt in this book will be beneficial for the student community, industrialists, researchers, their mentors, and to all people who wish to explore the applications of machine learning. We are greatly thankful to our contributors who hail from renowned institutes and industries that made a remarkable contribution by imparting their knowledge for the welfare of society. We express our sincere, wholehearted thanks to our editorial and production teams for their relentless contribution and for rendering unconditional support to publish this book on time.

Editors

Hemachandran K has been a passionate teacher for 14 years, with 5 years of research experience. He is a strong educational professional with a flair for science, highly skilled in artificial intelligence and machine learning. After earning a PhD in embedded systems at Dr. M.G.R. Educational and Research Institute, India, he started conducting interdisciplinary research in artificial intelligence. He is an open-minded and positive person who has stupendous peer-reviewed publication records with more than 20 journals and international conference publications. He served as an effective resource person at various national and international scientific conferences. He has a rich research experience in mentoring undergraduate and postgraduate projects. He holds two patents to his credentials. He has life membership in esteemed professional institutions. He was a pioneer in establish Single Board Computer lab at Ashoka Institutions, Hyderabad, India. Because of his self-paced learning schedule and thirst for upgrading and updating learning skills, he was awarded around 15 online certificate courses conferred by COURSERA and other online platforms. His editorial skills led him to be included as an editorial board member for numerous reputed SCOPUS/SCI journals.

Shubham Tayal is Assistant Professor in the Department of Electronics and Communication Engineering at SR University, Warangal, Telangana, India. He has more than 6 years of academic/research experience of teaching at the UG and PG levels. He earned a PhD in microelectronics and VLSI design at the National Institute of Technology, Kurukshetra; an MTech (VLSI design) at YMCA University of Science and Technology, Faridabad; and a BTech (electronics and communication engineering) at MDU, Rohtak. He has published more than 25 research papers in various international journals and conferences of repute, and many papers are under review. He is on the editorial and reviewer panels of many SCI/SCOPUS-indexed international journals and conferences. Currently, he is the editor or coeditor for six books with CRC Press (Taylor & Francis Group, USA). He acted as a keynote speaker and delivered professional talks on various forums. He is a member of various professional institutions such as IEEE, IRED, etc. He is on the advisory panel of many international conferences. He is a recipient of the Green ThinkerZ International Distinguished Young Researcher Award 2020. His research interests include simulation and modeling of multi-gate semiconductor devices, device-circuit codesign in digital/analogue domain, machine learning, and IoT.

Preetha Mary George is an accomplished academician with over 14 years of experience in teaching and 5 years in research. Currently, she works as Associate Professor in the Department of Physics, Dr. M.G.R. Educational and Research Institute. Her research was focused on ultrasonics and water quality monitoring systems. She earned a PhD at the same Institute in acoustical studies on industrially important biomolecules. She was a recipient of the MGR research award in 2016 and was recognized with a special recognition award in 2017. For her most recent tenure in research, she had nourished her passion to publish around two Scopus indexed journals. She has recently published a patent in water monitoring systems. Her zeal toward interdisciplinary research abetted her to bridge science to technology. Her public relation, communication, and interpersonal skills are commendable. She has a lifetime membership in various professional institutions.

Parveen Singla is Professor in the Electronics and Communication Engineering Department at Chandigarh Engineering College, Chandigarh Group of Colleges, Landran, Mohali, Punjab. He earned a BE with honors in electronics and communication engineering at Maharishi Dayanand University, Rohtak; a master's with honors in technology in electronics and communication engineering from Kurukshetra University, Kurukshetra; and a PhD in communication systems at IKG Punjab Technical University, Jalandhar, India. He has 17 years of experience in the field of teaching and research. He has published more than 35 papers in various reputed journals and national and international conferences. He also organized more than 30 technical events for the students to enhance their technical skills and received the Best International Technical Event Organiser Award. He is the guest editor of various reputed journals. His interest areas include drone technology, wireless networks, smart antenna, and soft computing.

Utku Kose earned a BS in computer education in 2008 at Gazi University, Turkey, as a faculty valedictorian. He earned an MS in the field of computer and datascience in 2010 at Afyon Kocatepe University, Turkey, and a PhD in computer engineering in 2017 at Selcuk University, Turkey. Between 2009 and 2011, he has worked as Research Assistant in Afyon Kocatepe University. Following this, he has also worked as Lecturer and Vice Director at Afyon Kocatepe University between 2011 and 2012, as Lecturer and Research Center Director at Usak University between 2012 and 2017, and as Assistant Professor at Suleyman Demirel University between 2017 and 2019. He is Associate Professor at Suleyman Demirel University, Turkey. He has more than 100 publications including articles, authored and edited books, proceedings, and reports. He is also in editorial boards of many scientific journals and serves as one of the editors of the Biomedical and Robotics Healthcare book series by CRC Press. His research interest includes artificial intelligence, machine ethics, artificial intelligence safety, biomedical applications, optimization, chaos theory, distance education, e-learning, computer education, and computer science.

Contributors

Melih Ağraz
Department of Applied Mathematics
Brown University
Providence, Rhode Island, USA

Mahesh Akuthota
Department of Computer Science and
    Engineering
Visvesvaraya College of Engineering and
    Technology
Hyderabad, India

Ezgi Ayyıldız
Turkish National Research Institute of
    Electronics and Cryptology (TÜBİTAK
    BİLGEM)
Gebze, Kocaeli, Turkey

Pragya Chandra
G.H. Raisoni Institute of Engineering and
    Technology
Nagpur, India

Sandip Kumar Chaurasiya
Department of Cybernetics
School of Computer Science
University of Petroleum and Energy
    Studies
Dehradun, India

Nilesh Deotale
G.H. Raisoni Institute of Engineering and
    Technology
Nagpur, India

Prathamesh Dherange
G.H. Raisoni Institute of Engineering and
    Technology
Nagpur, India

Akash Gurrala
School of Technology
Woxsen University
Hyderabad, India

Vasireddy Bindu Hasitha
School of Technology
Woxsen University,
Hyderabad, India

Syed Hasan Jafar
School of Business
Woxsen University
Hyderabad, India

Pokala Pranay Kumar
School of Business
Woxsen University
Hyderabad, India

Korupalli V Rajesh Kumar
School of Electronics Engineering
Vellore Institute of Technology
Chennai, India

M. Lavanya
Department of Electrical and Computer
    Engineering
Institute of Aeronautical Engineering
Hyderabad, India

Rajan Maduri
Vignan Institute of Science and Technology
Hyderabad, India

Saibaba V More
G.H. Raisoni Institute of Engineering and
    Technology
Nagpur, India

Mustafa Özgür Cingiz
Computer Engineering Department
Faculty of Engineering and Natural
    Sciences
Bursa Technical University
Bursa, Turkey

Vilda Purutçuoğlu
Department of Statistics
Middle East Technical University
Ankara, Turkey

H. Raghupathi
Department of Electrical and Computer
 Engineering
Visvesvaraya College of Engineering and
 Technology
Hyderabad, India

G Ravi
Department of Electrical and Computer
 Engineering
Visvesvaraya College of Engineering and
 Technology
Hyderabad, India

Eguturi Manjith Kumar Reddy
School of Technology
Woxsen University,
Hyderabad, India

M. Narendra Reddy
Department of Electrical and Computer
 Engineering
Institute of Aeronautical Engineering
Hyderabad, India

Pratiksha Repaswal
G.H. Raisoni Institute of Engineering and
 Technology
Nagpur, India

V Sailaja
Department of Electrical and Computer
 Engineering
Pragati Engineering College
Kakinada, India

M. Saritha
Department of Electrical and Computer
 Engineering
Institute of Aeronautical Engineering
Hyderabad, India

Deniz Seçilmiş
Department of Biochemistry and
 Biophysics
Stockholm University
Stockholm, Sweden

G S Sivakumar
Department of Electrical and Computer
 Engineering
Pragati Engineering College
Kakinada, India

K Sudhaman
Department of Electrical and Computer
 Engineering
Dr. MGR Educational and Research
 Institute University
Chennai, India
Hyderabad, India

P Suneetha
Department of Electrical and Computer
 Engineering
Pragati Engineering College
Kakinada, India

1

Introduction to Naive Bayes and a Review on Its Subtypes with Applications

Eguturi Manjith Kumar Reddy, Akash Gurrala, and Vasireddy Bindu Hasitha
Woxsen University

Korupalli V Rajesh Kumar
Vellore Institute of Technology

CONTENTS

DOI: 10.1201/9781003164265-1

1.1 Introduction

AI – Artificial Intelligence – is taking over the industrial, educational, medical, entertainment, and almost all sectors. In this aspect, machine learning (ML), deep learning, reinforcement learning, and natural language processing techniques, methods, and technologies have become more popular. All such technologies created an impact on real-time applications. With regard to ML applications, there are numerous functionalities and applications are revolutionizing all sectors. Generally, ML algorithms are broadly categorized into two major types:

1. Supervised learning algorithms
2. Unsupervised learning algorithms

Supervised learning is a type of learning in which data is provided to the algorithm with both input and output with labels.

Figure 1.1, shows the flow of the ML algorithm. Our focus is on supervised learning algorithms, mainly on classification models using the naive Bayes flow.

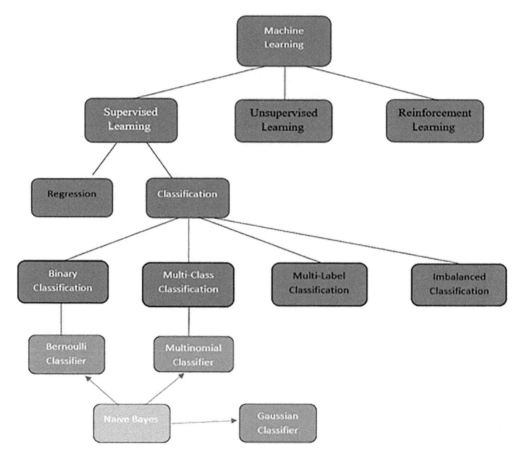

FIGURE 1.1
Machine learning algorithms – supervised learning.

The ML algorithm learns the process by setting a target. The machine will predict the new outcome to a newly furnished input from experience of past data. Mainly, learning systems are categorized into two under supervised learning models (Bhogaraju and Korupalli, 2020; Kumar et al., 2020; Bhogaraju et al., 2021; Seeja et al., 2021).

1. Regression modeling
2. Classification modeling

Classification and prediction are essential aspects of ML. Classification is a part of supervised learning, which categorizes the given data into classes. The data may be structured or unstructured. The algorithms in ML used for classification are called classification algorithms. These algorithms use trained data to predict the probability of data falling into the respective class. One of the most commonly used applications of classification is the sorting of standard emails and spam emails. In classification, the output is discrete. There are many algorithms used for classification, among which Naive Bayes is one of the most widely used simple and effective algorithm for classification.

1.2 Intuition behind the Naive Bayes Algorithm and Its Subtypes with Applications

Naive Bayes is a supervised learning algorithm, which is used as a classifier for ML and is based on the Bayes theorem. Naive Bayes works on probability distribution. The naive Bayes algorithm is one of the most influential and widely used classifiers, which helps build fast ML models that make rapid predictions with better accuracy.

1.2.1 Why Is It Called Naive Bayes?

In naive type, the features present in the data set are used to determine the outcome, these features do not interfere nor have any relation to other features. For example, let us consider a data set that contains the taste, size, color, and shape of a fruit. Depending on the data, the fruit is classified into apple or banana or mango. Here the features present in the data set do not depend on each other (Rish and others, 2001; Islam et al., 2007; Soria et al., 2011; Berrar, 2018).

1.2.2 Bayes Theorem – Intuition behind the Classification

1.2.2.1 Bayes Theorem

This theorem is just a simple extension of conditional probability where we have a formula for finding the probability of event A after event B has occurred.

$$\ggg P(A \mid B) = \frac{P(A \cap B)}{P(B)} \quad \text{and} \quad P(B \mid A) = \frac{P(A \mid B)P(B)}{P(A)}$$

From conditional probability

$$P(A \cap B) = P(A)P(B|A) \quad \text{and} \quad P(A \cap B) = P(B)P(A|B)$$

On equating the two equations above we get,

$$\text{The Bayes equation} \ggg P(A|B) = P(A)\frac{P(B|A)}{P(B)}$$

1.2.2.2 Bayes Theorem in Machine Learning

An A in the formula means the result or hypothesis and B implies any feature in the data or the evidence provided (Rish and others, 2001; Mukherjee and Sharma, 2012).

$$\text{Posterior} = \frac{(\text{Likelihood} * \text{Prior})}{\text{Normalization}}$$

Posterior – the probability of outcome or occurrence of result [$p(A|B)$] provided that it satisfies the feature(B) present in the dataset.

    Likelihood – $p(B|A)$ probability of the feature in data, given that hypothesis(A) should be true for the provided feature(B).

    Prior – probability of result or hypothesis that relates to all features of the data provided. Or simply the probability of an event before the data is seen.

    Normalization or marginal-pure probability of any feature given in the data set.

1.2.3 Types of Naive Bayes Models

1. Gaussian
2. Bernoulli
3. Multinomial

1.2.4 Gaussian Naive Bayes

Gaussian naive Bayes is just the augmentation of naive Bayes. Although there are other functions that may be used to estimate the data distribution, the Gaussian distribution could be the most convenient to work since it just requires one to calculate the descriptive statistics from the data sets. Gaussian naive Bayes can be excellently depicted by a bell-shaped curve (Gaussian Naive Bayes, 2020) (Figure 1.2).

    Gaussian distribution cannot be used with discrete data samples; it should only be used when the training data is continuous. In discrete data, we will only have a limited number of possible values. Continuous data is data that can take any value. Continuous data includes measurements such as weight, height, temperature, and volume. Naive Bayes can even be generalized to legitimate attributes by assuming a Gaussian distribution, which is the most general assumption. For other types of distributions, we need to calculate the probabilities for input values using a frequency, but in Gaussian distribution, we calculate standard deviation and mean in order for the distribution to be summarized.

    A simple way to calculate the mean is

$$\text{mean}(x) = 1 / n^* \text{sum}(x)$$

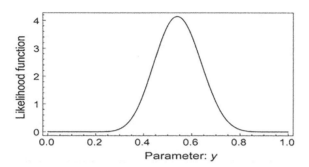

FIGURE 1.2
Bell-shaped curve, naive Bayes (Naive Bayes Classifiers - GeeksforGeeks, 2020).

In the above equation n represents the number of occurrences and x represents the different input values in the training data.

Similarly, we can calculate the standard deviation:

$$\text{Standard deviation}(x) = \text{sqrt}\left(1\,/\,n * \text{sum}\left(xi - \text{mean}(x)\hat{\ }2\right)\right)$$

(Naive Bayes for Machine Learning, 2020)

This is now the square root of its mean squared difference within each individual x value as well as the average of x values. xi is a unique x variable value for the ith case.

1.2.5 Predictions Using Gaussian Naive Bayes Model

Gaussian distribution can be excellently depicted by a bell-shaped curve. Using the Gaussian probability density function (PDF), we calculate the probabilities of new x values.

When making predictions, we integrate the above values into the Gaussian PDF along with a new variable input, and the Gaussian PDF returns an estimate of the probability of the new input value for that class.

$$\text{pdf}(x, \text{mean}, \text{sd}) = \left(1/\left(\text{sqrt}(2 * \text{PI}) * \text{sd}\right)\right) * \exp\left(-\left(\left(x - \text{mean}\hat{\ }2\right)/\left(2 * \text{sd}\hat{\ }2\right)\right)\right)$$

The probabilities can then be entered into the equation above to allow estimations for real-valued inputs. For instance, consider using the aforementioned computations, for predicting a car going out with a parameter of the weather condition (Naive Bayes for Machine Learning, 2020).

$$\text{go} - \text{out} = P\left(\text{pdf(weather)}|\text{class} = \text{go} - \text{out}\right) * P\left(\text{pdf(car)}|\text{class} = \text{go} - \text{out}\right) * P\left(\text{class} = \text{go} - \text{out}\right)$$

1.2.6 Bernoulli Classification

This type of classification classifies the outcome into only two features or two classes such as true or false, yes or no, present or absent, spam or not spam, male or female, and 0 or 1. So, when the outcome has to be predicted as a class type or a definite type with two variables, we use this type of classification. This classification works on Bernoulli distribution.

1.2.6.1 Bernoulli Statistics or Distribution

This type of distribution combines or relates probabilities and binary values 0 and 1 to obtain the probability of success and failure.

Let us consider p=probability of success and q=probability of failure $\gg q=1-p$

The formula that represents Bernoulli distribution is

$$P(X = x) = p^x(1-p)^{1-x}$$

here the values of x can only be 0 and 1, as follows:

The Bernoulli distribution

$$p(x) = P[X = x] = \begin{cases} q = 1-p & x = 0 \\ p & x = 1 \end{cases}$$

1.2.6.2 Rule for Bernoulli Naive Bayes Classifier

$$P\left(x_i|y\right) = P\left(i|y\right)x_i + \left(1 - P\left(i|y\right)\right)(1 - x_i)$$

Here i indicates the feature of the data set.

1.2.6.3 An Example for Bernoulli Naive Bayes

Let us consider a data set with categorical variables whose respective result is yes or no. This data consists of information regarding student confidence, whether they studied, and are they sick or not, and based on this, the information of whether they will pass the exam or not is provided (Table 1.1).

The probabilities are as follows: probability to pass=3/5 and probability to fail=1–3/5=2/5

Now to obtain the individual probabilities, we will calculate both the probabilities (pass and fail) for the given conditions of student results.

P(Confident = yes|Result = Pass) = 2/3
P(Studied = yes|Result=Pass)=2/3
P(Sick = yes|Result = Pass) = 1/3
P(Confident = yes|Result = Fail) = 1/2
P(Studied = yes|Result = Fail) = 1/2
$P(X|$Result = Pass$)\times P($Result = Pass$)$ = (2/3) * (2/3) * (1/3) * (3/5) = 0.088

TABLE 1.1

Student Result Dataset

Confident	Studied	Sick	Result
Yes	No	No	Fail
Yes	No	Yes	Pass
No	Yes	Yes	Fail
No	Yes	No	Pass
Yes	Yes	Yes	Pass

$P(X|\text{Result}=\text{Fail})\times P(\text{Result}=\text{Fail})=(1/2)*(1/2)*(1/2)*(2/5)=0.05$
Confident = yes, studied = yes, and sick = yes, then
$P(X)=P(\text{Confident}=\text{yes})\times P(\text{Studied}=\text{yes})\times P(\text{Sick}=\text{yes})=(3/5)*(3/5)*(3/5)=0.216$

And finally,

$$P\big(\text{Result}=\text{Pass}\,|\,X\big)=0.088/0.216=0.407$$

$$P\big(\text{Result}=\text{Fail}\,|\,X\big)=0.05/0.216=0.231$$

As the probability of pass is more than fail (0.407>0.231), the student will pass (Bernoulli Naive Bayes, 2020).

1.2.6.4 Advantages

- This type of classification is fast as it has to compute some simple probabilities.
- In this classification, every feature variable is independent >> one event does not depend on other events.
- Every variable is stored as 0 and 1 or yes or no, which makes this classification easier.
- If the provided data set is small, then the predicted results will be more accurate.

1.2.6.5 Disadvantages

- Sometimes the features in the data set may be dependent. By this, the actual outcome might not match the predicted outcome.

A naive classifier sometimes makes some strong assumptions based on experience with the training data. Sometimes, this assumption will not match the actual outcome.

1.2.7 Multinomial Naive Bayes Classifier

Generally, this classifier is most commonly used in natural language processing (NLP) as a probabilistic learning method. This is especially used when the data consists of words. This classifier has a higher success rate in document classification than any other algorithm (Liu and others, 2010; Moghaddam and Ester, 2011; Martin and Johnson, 2015; Sun, Huang and Qiu, 2019; Uzun, 2020) (Table 1.2).
    Example: Let us Consider the Following Data given in the Table Below
    Here, we are required to choose a class for the test document.

1. $P(c)=\dfrac{N(c)}{N}$

where, $P(c)$ is probability of class
    $N(c)$ is the number of documents with that class
    N is the total number of documents.

TABLE 1.2

Given Data

Phases	Document	Words	Class
Training	1	India Africa India	r
	2	India India America	r
	3	India Australia	r
	4	England Japan India	s
Test	5	India India India England Japan	?

2. $P(w|c) = \dfrac{\text{count}(w,c)+1}{\text{count}(c)+|V|}$

where, $P(w|c)$ is the probability of word given a class
count (w,c) is the count of the word that occurs in that class
count(c) is the count of all words in that class
Priors:

$$P(r) = \frac{3}{4}$$

$$P(s) = \frac{1}{4}$$

Conditional probabilities of the testing documents:

$$P(\text{India}\,|\,r) = \frac{(5+1)}{(8+6)} = \frac{6}{14} = \frac{3}{7}$$

$$P(\text{England}\,|\,r) = \frac{(0+1)}{(8+6)} = \frac{1}{14}$$

$$P(\text{Japan}\,|\,r) = \frac{(0+1)}{(8+6)} = \frac{1}{14}$$

$$P(\text{India}\,|\,s) = \frac{(1+1)}{(3+6)} = \frac{2}{9}$$

$$P(\text{England}\,|\,s) = \frac{(1+1)}{(3+6)} = \frac{2}{9}$$

$$P(\text{Japan}\,|\,s) = \frac{(1+1)}{(3+6)} = \frac{2}{9}$$

Choosing a class:

$$P(r\mid d5)a\frac{3}{4}*\left(\frac{3}{7}\right)^{3}*\frac{1}{14}*\frac{1}{14}$$

$$\approx 0.0003$$

$$P(s\mid d5)a\frac{1}{4}*\left(\frac{2}{9}\right)^{3}*\frac{2}{9}*\frac{2}{9}$$

$$\approx 0.0001$$

Therefore, $P(r\mid d5)>P(s\mid d5)$.
  Hence, the tested data is classified into class "*r*."

1.2.8 Differences between Gaussian, Bernoulli, and Multinomial Distributions

Continuous distribution was what Gaussian was founded on. Bernoulli, on the other hand, is useful in determining whether a feature is present or not. Finally, multinomial naive Bayes considers an attribute in which a given word denotes the number of times it appears or how often it occurs (Multinomial Naive Bayes Explained, 2020).

1.2.9 Advantages of Naive Bayes

- It can work effectively with large datasets, and for small datasets, it can perform the most effective alternatives.
- It is quick, simple to execute, and has a higher accuracy in prediction.
- It is much helpful when it comes to text classification and can also work with multiclass prediction problems.

1.2.10 Disadvantages of Naive Bayes

- It make it hard to obtain the set of independent variables for developing a model (Multinomial Naive Bayes Explained, 2020).

1.3 Real-Time Application: Human Activity Recognition Using Naive Bayes Algorithm

Research on human activities recognition (HAR) is increasing day by day, and modeling aspects are also advancing. Researchers' interest in HAR changes the model's potentiality and to helps developing innovative methods. Various ML algorithms can be useful to classify HAR, but here we have considered naive Bayes algorithm to find the hypothesis behind the activity recognition process.

The Human Activities Dataset was downloaded from the UCI Repository web source, an open-source database. Here, the dataset consists of the daily activities of 30 subjects, data collected from waist-mounted smartphones, with embedded inertial measurement sensors. The objective is to classify the activities based on performed activities (Anguita et al., 2012; Anguita, Ghio, Oneto, Parra and Reyes-Ortiz, 2013; Anguita et al., 2013; Reyes-Ortiz et al., 2013).

Figure 1.3, shows the six different activities performed by the 30 subjects, those who wore a smartphone on their waist during data collection. These 30 subjects performed these tasks in a disciplined manner. The mechanism of the preprocessing techniques well explained the research works. Here, the intention was to show the performance of the naive Bayes algorithm. Next we see the results obtained with the naive Bayes algorithm with comparative analysis between a few more supervised algorithms.

1.3.1 Dataset Attributes

- Tri-axial Acceleration (Body Acceleration) – X_a, Y_a, Z_a
- Gyroscope – Angular Velocity - X_g, Y_g, Z_g
- Time and Frequency Domain Feature Variables
- Activity Labels – Target Variable

1.3.2 Naive Bayes Algorithm–Based Result

Figure 1.4 shows a confusion matrix based on the naive bayes classification algorithm, with approximately 99% accuracy. The performance matrix of the model shows the significant results. The results of the comparison between the naive Bayes algorithm and other supervised learning models are shown in Figure 1.5.

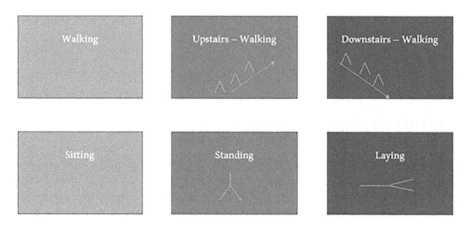

FIGURE 1.3
HAR – six different activities.

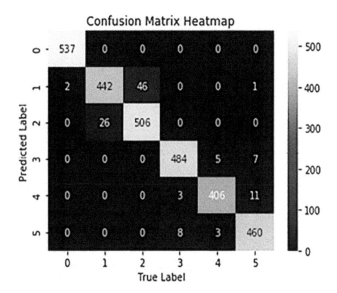

FIGURE 1.4
Confusion matrix of activity recognition – using naive Bayes.

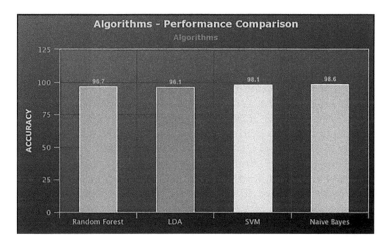

FIGURE 1.5
Algorithms – performance comparison with HAR dataset.

1.4 Conclusion

This conceptual explanation of the naive Bayes algorithm along with detailed follow-up with applications clearly signifies the intuition behind ML-based supervised learning models. Mainly the mathematical properties of naive Bayes and Bayes theorem were explained clearly. Applications like the HAR were explained, with a comparative analysis between supervised learning algorithms. This chapter will help students learn about the naive Bayes theorem and its applications.

References

Anguita, D. et al. (2012) 'Human activity recognition on smartphones using a multiclass hardware-friendly support vector machine', in *International workshop on ambient assisted living*, pp. 216–223.

Anguita, D., Ghio, A., Oneto, L., Parra, X., Reyes-Ortiz, J. L., et al. (2013) 'A public domain dataset for human activity recognition using smartphones', in *Esann*, p. 3.

Anguita, D., Ghio, A., Oneto, L., Parra, X. and Reyes-Ortiz, J. L. (2013) 'Energy efficient smartphone-based activity recognition using fixed-point arithmetic', *Journal of Universal Computer Science*, 19(9), pp. 1295–1314.

Berrar, D. (2018) 'Bayes' theorem and naive Bayes classifier', *Encyclopedia of Bioinformatics and Computational Biology: ABC of Bioinformatics*; Elsevier Science Publisher: Amsterdam, The Netherlands, pp. 403–412.

Bhogaraju, S. D. et al. (2021) 'Advanced Predictive Analytics for Control of Industrial Automation Process', in *Innovations in the Industrial Internet of Things (IIoT) and Smart Factory*. IGI Global, pp. 33–49.

Bhogaraju, S. D. and Korupalli, V. R. K. (2020) 'Design of smart roads-a vision on indian smart infrastructure development', in *2020 International Conference on COMmunication Systems \& NETworkS (COMSNETS)*, pp. 773–778.

Gaussian Naive Bayes (2020).

Islam, M. J. et al. (2007) 'Investigating the performance of naive-bayes classifiers and k-nearest neighbor classifiers', in *2007 International Conference on Convergence Information Technology (ICCIT 2007)*, pp. 1541–1546.

Kumar, K. V. R. et al. (2020) 'Internet of things and fog computing applications in intelligent transportation systems', in *Architecture and Security Issues in Fog Computing Applications*. IGI Global, pp. 131–150.

Liu, B. and others (2010) 'Sentiment analysis and subjectivity.', *Handbook of Natural Language Processing*, 2(2010), pp. 627–666.

Martin, F. and Johnson, M. (2015) 'More efficient topic modelling through a noun only approach', in *Proceedings of the Australasian Language Technology Association Workshop 2015*, pp. 111–115.

Moghaddam, S. and Ester, M. (2011) 'ILDA: interdependent LDA model for learning latent aspects and their ratings from online product reviews', in *Proceedings of the 34th International ACM SIGIR Conference on Research and Development in Information Retrieval*, pp. 665–674.

Mukherjee, S. and Sharma, N. (2012) 'Intrusion detection using naive Bayes classifier with feature reduction', *Procedia Technology*, 4, pp. 119–128.

Multinomial Naive Bayes Explained (2020) *Great Learning Team*. https://www.mygreatlearning.com/blog/multinomial-naive-bayes-explained/

Naive Bayes Classifiers - GeeksforGeeks (2020) *GeeksforGeeks*. https://www.geeksforgeeks.org/naive-bayes-classifiers/

Naive Bayes for Machine Learning (2020) *Machine Learning Matery*. https://machinelearningmastery.com/naive-bayes-for-machine-learning/

Reyes-Ortiz, J. L. et al. (2013) 'Human activity and motion disorder recognition: towards smarter interactive cognitive environments', in *ESANN*.

Rish, I. and others (2001) 'An empirical study of the naive Bayes classifier', in *IJCAI 2001 workshop on empirical Methods in Artificial Intelligence*, pp. 41–46.

Seeja, G. et al. (2021) 'Internet of things and robotic applications in the industrial automation process', in *Innovations in the Industrial Internet of Things (IIoT) and Smart Factory*. IGI Global, pp. 50–64.

Singh, G., Kumar, B., Gaur, L., and Tyagi, A. (2019, April). 'Comparison between multinomial and Bernoulli naïve Bayes for text classification', in *2019 International Conference on Automation, Computational and Technology Management (ICACTM)*, pp. 593–596.

Soria, D. et al. (2011) 'A 'non-parametric'version of the naive Bayes classifier', *Knowledge-Based Systems*, 24(6), pp. 775–784.

Sun, C., Huang, L. and Qiu, X. (2019) 'Utilizing BERT for aspect-based sentiment analysis via constructing auxiliary sentence', *arXiv preprint arXiv:1903.09588*.

Uzun, E. (2020) 'A novel web scraping approach using the additional information obtained from web pages', *IEEE Access*, 8, pp. 61726–61740.

2

A Review on the Different Regression Analysis in Supervised Learning

K Sudhaman

Dr. MGR Educational and Research Institute University

Mahesh Akuthota

Visvesvaraya College of Engineering and Technology

Sandip Kumar Chaurasiya

School of Computer Science, University of Petroleum and Energy Studies

CONTENTS

2.1 Introduction

In supervised learning, data is provided to the machine learning system with both input and output. Through the experience of past data, the machine will be able to predict the new outcome to a newly provided input. In this process, regression is a stage where the machine will be able to figure out (or) establish a relation (or) function between input and the outcome provided in the data set. And by using this relation or the function, the machine will take the input and run it through the function and get the possible outcome.

Generally, regressions contain two types of variables: dependent and independent; there might be multiple independent variables, such as the different characteristics of the input

DOI: 10.1201/9781003164265-2

data set, and one dependent variable, that is, the outcome or the result. Regression helps us to understand which factors (variables) are important and which are not and can be neglected in the relation or equation, which helps speed up the processing.

For example, let us consider a data set on athletes, containing their weight, height, stamina, time of practice infield, muscle strength, and endurance as input, and their like track record as output. From regression, we obtain a function between independent variables (weight, height, time of practice, etc.) and dependent variable (races they have won). Suppose a new competitor enters the race; the machine will take their weight, stamina, time of practice, etc. as input and substitute it in the function obtained. From this, we will be able to decide whether they will win the race or not.

Suppose $Y = f(x_1, x_2, x_3, x_4 \ldots)$ is the regression or the relation.

where Y (dependent variable) is the outcome that determines the performance of the player, x (independent variable) terms are related to the characteristics of the player, and $f(x)$ is the function between x and y

Every time a new player competes, their characters stored as variables x will be substituted in the function, and by this, their probability of winning can also be predicted.

There are several types of regressions:

A. Linear regression
B. Logistic regression
C. Regularization
    a. Ridge Regression
    b. Lasso Regression
D. Polynomial regression
E. Bayesian regression

2.2 Linear Regression

The word itself gives an idea about this type of regression; it simply means that the relation between the dependent (outcome) and independent (input) variable is a linear equation, as shown in Figure 2.1. Linear regression is of two types: single independent variable

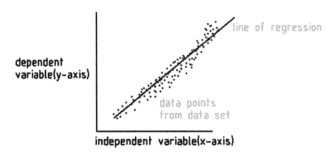

FIGURE 2.1
Linear regression graph.

and multivariable, which simply means that the outcome depends on two or more characteristics of the input data.

2.2.1 Simple Linear Regression

In this type, the dependent variable depends only on a single independent variable. The regression is

$$Y = mX + c + e$$

where e is the calculation error, c is the intercept, and m is the slope of the equation. Generally, when a relation is plotted, all the points in the data set cannot be assigned to a single relation; in the graph there are some points above and below the straight line, and to accommodate this, an extra e is added.

2.2.2 Finding the Line Equation for the Data

To find a line that fits the data we first need to learn some terminologies; residual is the distance from a point to the line. First consider a line parallel to the x-axis then find the residuals for all the points, square them and sum them up, next rotate the line at a small angle and then again find the sum of the squares of the residuals, continue this process till you get the minimum value of the sum of the squares of the residuals. The line with the minimum sum will fit the data best.

We will not be able to get the exact line that fits the data through this trial-and-error method, so first plot a graph between the sum of the squares of the residuals on the y-axis and the respective line on the x-axis, as shown in Figure 2.2.

We find the point at which the derivative is 0, and that line is the one that fits the data. The lines on the x-axis just differ in slope and intercept values. We can also use a 3D graph in which we can have both intercept and slope values on different axes, as shown in Figure 2.3.

$$Y = \beta_1 + \beta_0 x \tag{2.1}$$

Then, after solving, we get

$$\beta_1 = \frac{\sum_{i=1}^{n}(x_i - x)(y_i - y)}{\sum_{i=1}^{n}(x_i - x')^2} \tag{2.2}$$

$$\beta_0 = y' - \beta_1 x' \tag{2.3}$$

(*Linear Regression in Machine learning - Javatpoint*, 2020).

where $\bar{x}$ and $\bar{y}$ indicate mean. After obtaining the intercept and the slope we also need to check that the values obtained will be able to predict the relevant value or not, for which we need to find the value of p. The value p can justify the significance of the coefficient in predicting the targeted outcome, as a general condition or observed result is that if the p-value is less than 0.05, then the obtained coefficients strongly satisfies the relation between dependent and independent variable.

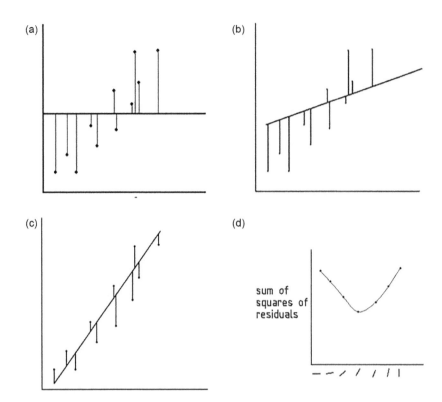

FIGURE 2.2
(a)–(d) Procedure to find the best fit line.

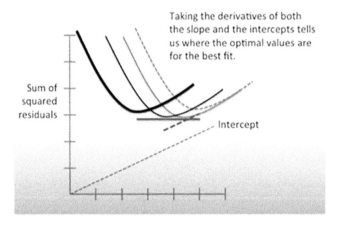

FIGURE 2.3
3D graph.

Now we know that the coefficients are satisfactory to the relation; we also need to know how well the combination of coefficients and independent variable or simply the function is satisfying the relation, or if it is not.

For that, we use residual standard error (RSE) and R^2 statistics

$$RSE = \sqrt{\frac{1}{n-2} \, RSS} = \sqrt{\frac{1}{n-2} \sum_{i=1}^{n} (y_i - y')^2} \tag{2.4}$$

$$R^2 = \frac{TSS - RSS}{TSS} = 1 - \frac{RSS}{TSS}, \; TSS = \sum (y_i - y')^2 \tag{2.5}$$

(*The Complete Guide to Linear Regression in Python | by Marco Peixeiro | Towards Data Science,* 2020).

$$R^2 = (\text{variation of data with mean line}) - (\text{variation of data with regression line}) /$$

$$(\text{variation with mean line})$$

R^2 is ranged between 0 and 1, the lesser the variation along the line, the more efficient regression works. R^2 closer to 1 implies the function is highly satisfying the relation, or else not.

For example, a man goes to a grocery store regularly and buys apples (note: the quality of apples is the same all the time) depending on the price; the number of apples bought will vary, which means if the price is high he will buy fewer apples and if the price is less he will buy more apples. Depending on his past 6 months purchase data we plot a graph with the price of apples (independent variable) on the x-axis and number of apples bought (dependent variable) on the y-axis and then we obtain a linear equation with a negative slope because the number of apples bought will decrease as the price increases, as shown in Figure 2.4.

From this equation, we will be able to predict how many apples he will buy next time he goes to the grocery store.

2.2.3 Multivariable Linear Regression

In this type of regression, the dependent variable will depend on two or more independent variables. Depending on the variable coefficients we can decide what factor from the input data list will be more effective on the dependent variable. In simple linear regression, we fit a line to the data the same way we fit a plane or a higher-dimension object (adding additional variables) to the data in multivariable linear regression, calculating R^2 and p

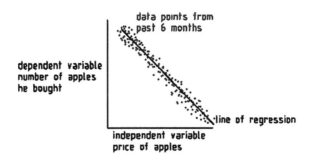

FIGURE 2.4
Prediction.

in the same way for both single variable and multivariable linear regression. In the case of multivariable regression, we use an extra term f. Generally, p is calculated to a specific coefficient only but in multivariable regression, we have more independent variables so f will describe how well the combination of coefficients works.

$$Y = \beta_0 + \beta_1 X_1 + \beta_2 X_2 + \cdots + \beta_p X_p \tag{2.6}$$

$$F = \frac{\dfrac{TSS - RSS}{p}}{\dfrac{RSS}{(n - p - 1)}} \tag{2.7}$$

(The Complete Guide to Linear Regression in Python | by Marco Peixeiro | Towards Data Science, 2020).

If the value of f is high this indicates that there is a strong relationship between the dependent and independent variables. Otherwise, it will be approximately equal to 1. Generally, if the dependent variable depends on more number independent variables then calculating p for each variable coefficient is not that efficient, and there might be chances that some variable coefficient with a lower p-value will be less effective, or sometimes have no effect on the output because of these variables. So, f is used to neglect these non-effective variables and should be calculated only once.

For example, we take a class of students and specify some characters such as their intelligence (IQ), time spent in problem-solving, amount of hard work done, attendance, roll number, their appearance, presenting skills, and their soft skills. And the dependent variable will give information on whether they pass the exam or not. On the bases of the given data, we obtain a multivariable linear equation, such as $y = a_0 + a_1 x_1 + a_2 x_2 + a_3 x_3 + a_4 x_4 + \ldots + e$.

The coefficient for variables such as the amount of hard work and intelligence will be high and for variables such as attendance, roll number, and their appearance the coefficient will be low and sometimes can be neglected because these characteristics do not have much effect on how they write their exam. Even if a new student enters the class based on their abilities we will be able to decide if they pass the exam or not. By this type of regression, we can identify things that affect the outcome more and be able to improve them for better results.

2.3 Logistic Regression

First, we need to understand why logistic regression is used instead of linear regression. Suppose a data set with both dependent and independent variables is given and generally the dependent variable in linear regression is continuous, that means there will always be a relevant outcome to the input provided in the data set, but in some situations the dependent variable will be categorical, simply yes/no or true/false, and in such problem statements the dependent variable cannot be determined from the independent variable using linear regression. This means the two variables cannot be related linearly. In this type of data set the outcome value is a probabilistic value that cannot exceed 1. For this type of regression we need a probabilistic relationship between the variables. Or, simply, we need a function that converts independent variables into the expression of probability. In logistic regression, the predicted outcome will never exceed 1 and will never go below 0.

In linear regression, we use a straight line as best fit but in logistic regression, we fit an s-shaped curve into the data set, which predicts a maximum of two values. The predicted outcome will be a probabilistic valve between 0 and 1, so to classify these values we consider a threshold value of 0.5, and points above the threshold value are considered to have the most possible outcome, or they are considered as *yes* or *true* type of categorical declarations. Points below the threshold value are considered as *no* or *false* type of predictions (*Logistic Regression in Machine Learning - Javatpoint*, 2020).

2.3.1 Logistic Function (Sigmoid Function)

The sigmoid function is a function that relates the variables in a probabilistic way; it ranges all the independent variable values between 0 and 1 of the dependent variable. On classifying the values based on threshold values, outcomes above 0.5 tend to 1 and outcomes below 0.5 tend to 0 (Figure 2.5).

2.3.2 Logistic Regression Equation

By using the equation of linear regression, we can derive a sigmoid function that can be used for the logistic regression.

Linear regression equation

$$Y = \beta_0 + \beta_1 X_1 + \beta_2 X_2 + \cdots + \beta_P X_P \tag{2.8}$$

Sigmoid function

$$P = \frac{1}{1 + e^{-y}} \tag{2.9}$$

Using y in the sigmoid function, we get the following result:

$$\ln\left(\frac{P}{1-P}\right) = \beta_0 + \beta_1 X_1 + \beta_2 X_2 + \cdots + \beta_P X_P \tag{2.10}$$

(*machine learning - Why is logistic regression called regression? - Stack Overflow*, 2020).

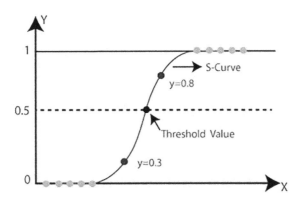

FIGURE 2.5
Sigmoid function graph.

2.3.3 Types of Logistic Regressions

Based on the number and types of categorical variables, logistic regression can be classified into three types:

Binomial: In binomial regression, the number of predicted outcomes will be only two, such as 0 or 1, true or false, yes or no, and there will be only two dependent variables.

Multinomial: In this type of regression there are three dependent variables, divided according to different categories, such fish, crab, and wales.

Ordinal: In ordinal regression, there are more than three dependent variables, such as in a pet category we can have dogs, cats, birds, rabbits, etc.

Logistic regression is used in detecting spam emails; this type of regression is a binomial logistic function, and the outcome here is just spam or not spam. To detect that, the machine will take several factors such as sender of the email, the number of typos in the email, and occurrence of words such as (offer, prize, gift, lottery) as independent variables and develop a sigmoid function based on the training data, and finally a logistic regression is obtained, which will be able to detect spam mails (*4 Logistic Regressions Examples to Help You Understand - Magoosh Data Science Blog*, 2018).

2.4 Regularization

Overfitting of data is a prevalent issue within the field of machine learning. To prevent this issue, we tend to use a process called regularization, which uses a penalty term in a multilinear regression model to avoid data overfitting, particularly where the trained and tested data are vastly different.

There are two main techniques in regularization:

  i. Ridge regression (L_2)
  ii. Lasso regression (L_1)

2.4.1 Ridge Regression (L_2)

Ridge regression is a regularization technique in which we will be adding a penalty term in a multilinear regression model, which is obtained from trained data. The expected outcome for the testing data would be more reliable, but the trained data will be less precise.

The general equation for a multilinear regression is

$$y = B_0 + B_1 x_1 + B_2 x_2 + \cdots + B_k x_k \tag{2.11}$$

where $x_1, x_2 \ldots x_k$ are independent variables and y is the dependent (output) variable. B is the regression coefficient.

λ (slope)^2 is a penalty term that we tend to add for ridge regression.

The equation for ridge regression is given below:

$$y = B_0 + B_1 x_1 + B_2 x_2 + \cdots + B_k x_k + \lambda (\text{slope})^2 \tag{2.12}$$

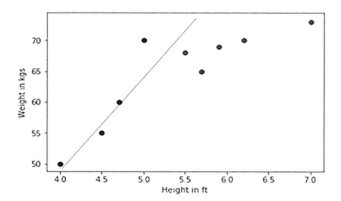

FIGURE 2.6
For training data, there is less variance, whereas for testing data, there is more variance.

where λ is the regularization parameter or degree of deflection from the original curve (Tarun 2019).

Ridge regression can shrink the slope close to 0 but not exactly 0. In contrast, ridge regression does greater when most variables are useful.

For example, let us take the weight of humans on the x-axis and their height on the y-axis, where blue dots represent trained data and red dots represent testing data. When we look at Figure 2.6, we can observe that the trained data and testing data vary largely.

If we obtain a linear regression for the trained data, we get the blue line as the best fit. But if we observe the line, it doesn't coincide with the testing data. Variance for the testing data and the regression line is much high. This means that if the testing data is varies a lot from trained data, the obtained regression will be not able to predict the outcome with more accuracy. To correct this issue, we tend to add a penalty term, i.e., λ **(slope)^2**.

When we take the $\lambda=1$, the graph in Figure 2.6 appears as the one in Figure 2.7.

The newly obtained regression line is known as the ridge regression line, which best fits the testing data and loses its accuracy to the trained data.

An increase in the λ value causes the slope of the ridge regression to reduce and become more horizontal. The model becomes less sensitive to the variations of the independent variable.

Ridge regression's drawbacks: Ridge regression reduces the complexity in parameters by reducing the slope towards a value that is similar to 0 but just not exactly 0. Therefore, this model isn't effective in reducing features. ("Lasso vs Ridge vs Elastic Net | ML - GeeksforGeeks" 2020).

2.4.2 Lasso Regression (L_1)

Lasso (least absolute shrinkage and selective operator) regression is similar to ridge regression.

It is one of the regularization techniques that introduces a bias term, i.e., the absolute value of the slope instead of squaring the slope.

The equation for lasso regression is given below:

$$y = B_0 + B_1 x_1 + B_2 x_2 + \cdots + B_k x_k + \lambda |\text{slope}|^2 \qquad (2.13)$$

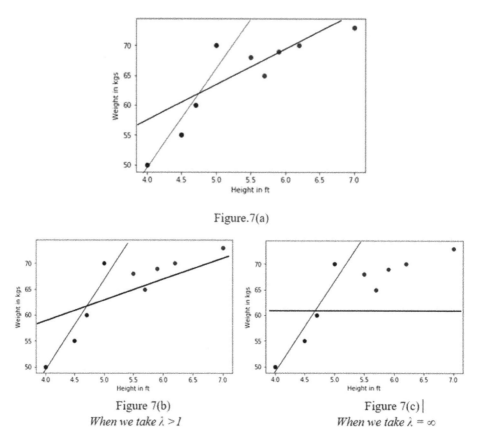

Figure.7(a)

Figure 7(b)
When we take λ > 1

Figure 7(c)|
When we take λ = ∞

FIGURE 2.7
(a)–(c) Ridge regression with different lambda values.

The variable y is the output parameter and x_1, x_2... x_k are the input variables, which are independent variables. B is the regression coefficient.

Here, λ is the regularization parameter or degree of deflection from the original curve (Tarun 2019).

Lasso regression can shrink the slope to be precisely equal to 0. It seems to be slightly faster compared to ridge regression when minimizing uncertainty throughout models with a bunch of unnecessary parameters because this could turn them away formulas. This makes the final equation simpler and easier to interpret. When lambda increases, the bias will increase. In the same way, if lambda decreases, variance will increase.

The main problem with lasso regression is that it holds on to only one variable when we have related variables and sets other related variables to 0, which will lead to some loss of data, this results in poor accuracy in our model ("What Is LASSO Regression Definition, Examples and Techniques" 2020).

Similar to the ridge regression, the slope of lasso regression line depends on the λ value.

Lasso regression improves the performance of estimation. If the interaction among the outcome factor and the determinant variables is roughly linear and we provide more tests, OLS (ordinary least squares) regression estimations would include fewer biases and variances. However, the variance of the OLS parameter estimates will be higher if there are comparatively fewer number observations and many predictors. In this case, shrinking the regression coefficients can reduce variance without a considerable increase in bias, which makes lasso regression helpful.

Model evaluation could be improved using lasso regression. In such an OLS multiple regression, the minimum value of a certain response variable always does not appear to be associated with the outcome variables. As a consequence, we often eventually wind up having an excessively fitting model, which is difficult to understand. The regression coefficients regarding insignificant parameters are reduced significantly throughout the lasso regression, essentially removing it throughout the model as well as making it even more comparable, allowing this to pick just the same recent predictor variables.

For example, let us solve the task of predicting the average meal price of a restaurant. The data consists of the following features.

Training set records: 13,984

Testing set records: 5,648

Features provided in the data are title, restaurant id, cuisines, opening/closing times, locality, rating, number of votes received, cost, and food wasted. After executing all the steps till feature scaling (excluding), we can proceed with building a lasso regression. Lasso regression excludes variables such as restaurant id and opening/closing times because these variables may not help us predict the price of an average meal, and excluding them helps us predict the required outcome more efficiently and more accurately ("What Is LASSO Regression Definition, Examples and Techniques" 2020).

2.4.3 Lasso Regression's Drawbacks

- Lasso has certain issues with certain sources of data. Unless the sample size (n) becomes less than the predictor variables (p), lasso regression would select within most n predictor variables to non-zero, although some certain predictor variables were important.
- If there are two or more strongly collinear variables, lasso regression will choose one at random, which is not ideal for data interpretation." ("Lasso vs Ridge vs Elastic Net | ML - GeeksforGeeks" 2020).

In both regressions

- The term lambda represents the sum of deformation.
- Lambda=0 indicates whether certain characteristics were considered; this also becomes similar to linear regression, which uses just the residual number of squares to construct a statistical model.
- Lambda=∞ denotes that no function was taken into account, i.e., when λ approaches infinity, this loses ever more characteristics.
- If lambda rises, the bias rises.
- As lambda is reduced, variance improves ("What Is LASSO Regression Definition, Examples and Techniques" 2020).

2.5 Polynomial Regression

In polynomial regression, the relation between the continuous parameter (y) and the predictor parameter (x) is designed as something of an nth dimension polynomial. It deals with a nonlinear data set using a linear model. It is tantamount to multiple linear regression. When the linear regression model is unable to capture the pattern of nonlinear data set, the problem of underfitting arises, where an underfit machine learning model will have underperformed on the training data. To avoid this, an underfit polynomial regression is used for nonlinear data which fits the nonlinear relationship between the value of x, which is the independent variable and the values of the dependent variables of y, which is the target variable more precisely. ("Machine Learning Polynomial Regression - Javatpoint" 2020) (Figure 2.8).

The polynomial regression equation is given as follows:

$$y = b_0 + b_1 x_1 + b_2 x_1^2 + \ldots b_n x_1^n \tag{2.14}$$

The independent parameter becomes x and the target output becomes y. The regression results are $b_0, b_1, b_2 \ldots, b_n$.

When we compare linear regression with polynomial regression, the above data points are nonlinear, so if we avail a linear model, we can see that it hardly nudges any

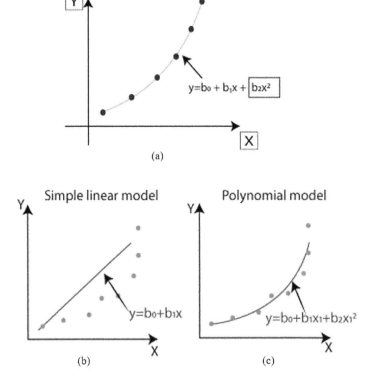

FIGURE 2.8
(a)–(d) Polynomial Regression (Adapted from "Machine Learning Polynomial Regression - Javatpoint" 2020)

point, whereas on the other hand polynomial model can cover many of the data points. Polynomial regression is also called polynomial linear regression as it does not depend on the variable whereas it depends on the regression coefficients.

Equations:

- Simple linear regression: $y=b_0+b_1x$
- Multiple linear regression: $y=b_0+b_1x_1+b_2x_2 \cdots + b_nx_n$
- Polynomial regression: $y=b_0+b_1x_1+b_2x_1^2 +\cdots\ b_nx_1^n$
  Adapted from ("Machine Learning Polynomial Regression - Javatpoint" 2020).
  Given n points $(x_1,y_1),(x_2,y_2)...(x_n,y_n)$, the best fit is $y=a_0+a_1x + a_2x^2$ to the data $(n \geq 3)$. Sum of the squares of the residuals is indeed the difference between what was observed and what was expected.

$$S_r = \sum_{i=1}^{n} \left(y_i - \left(a_0 + a_1x + a_2x^2 \right) \right)^2 \tag{2.15}$$

$$S_r = \sum_{i=1}^{n} \left(y_i - a_0 - a_1x - a_2x^2 \right)^2 \tag{2.16}$$

$$\frac{\delta s_r}{\delta a_0} = 0; \frac{\delta s_r}{\delta a_1} = 0; \frac{\delta s_r}{\delta a_2} = 0 \tag{2.17}$$

$$\frac{\delta s_r}{\delta a_0} = \sum_{i=1}^{n} 2 \left(y_i - a_0 - a_1x_i - a_2x_i^2 \right)(-1) = 0 \tag{2.18}$$

$$\frac{\delta s_r}{\delta a_1} = \sum_{i=1}^{n} 2 \left(y_i - a_0 - a_1x_i - a_2x_i^2 \right)(-x_i) = 0 \tag{2.19}$$

$$\frac{\delta s_r}{\delta a_2} = \sum_{i=1}^{n} 2 \left(y_i - a_0 - a_1x_i - a_2x_i^2 \right)\left(-x_i^2\right) = 0 \tag{2.20}$$

From these equations we get $y=a_0+a_1x+a_2x^2$ that gives us the regression model
Example:

- We recorded 18 vehicles as they passed through a tollbooth. We recorded each car's speed as well as the time of the day (hour) it passed us. The x-axis depicts the number of hours in the day, while the y-axis depicts the pace (Figure 2.9).
- Imported NumPy and matplotlib, draw the Polynomial Regression graph (Figure 2.10).
- Made the arrays that display the x- and y-axis values.
- Then determined where the line would appear, starting at position 1 and ending at position 22.
- The scattered plot of polynomial regression was the output.

```
import matplotlib.pyplot as plt
x = [1,2,3,5,6,7,8,9,10,12,13,14,15,16,18,19,21,22]
y = [100,90,80,60,60,55,60,65,70,70,75,76,78,79,90,99,99,100]

plt.scatter(x,y)
plt.show()
```

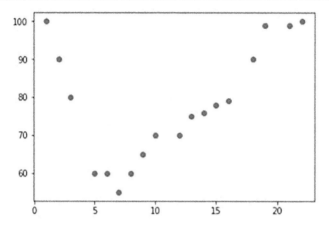

FIGURE 2.9
Scatter plot.

```
import numpy
x = [1,2,3,5,6,7,8,9,10,12,13,14,15,16,18,19,21,22]
y = [100,90,80,60,60,55,60,65,70,70,75,76,78,79,90,99,99,100]

mymodel = numpy.poly1d(numpy.polyfit(x,y,3))

myline = numpy.linspace(1,22,100)

plt.scatter(x,y)
plt.plot(myline,mymodel(myline))
plt.show()
```

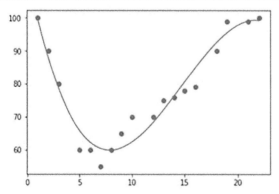

FIGURE 2.10
Polynomial regression graph.

```
import numpy
x = [1,2,3,5,6,7,8,9,10,12,13,14,15,16,18,19,21,22]
y = [100,90,80,60,60,55,60,65,70,70,75,76,78,79,90,99,99,100]

mymodel = numpy.poly1d(numpy.polyfit(x,y,15))

myline = numpy.linspace(1,22,100)

plt.scatter(x,y)
plt.plot(myline,mymodel(myline))
plt.show()
```

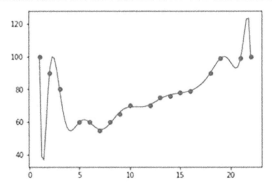

FIGURE 2.11
Polynomial graph with degrees of freedom.

If we change the degrees of freedom from 3 to 15, then the plot will be the best fit (Figure 2.11).

2.6 Bayesian Regression

The Bayesian method is a strategy to delineate and reckon statistical models. Bayesian regression is used when the data set has insufficient data and this method of regression is very useful when the given data is poorly distributed. When we train the machine learning model with the data set. As opposed to standard regression methods, the output of a Bayesian regression model is derived from a probability distribution, while the output of a regular regression model is merely obtained from one value of each attribute. A typical distribution is used to produce the output y (where the variance and mean have been standardized). Bayesian linear regression aims to find the model parameters' "posterior" distribution, rather than the model parameters themselves. The model parameters, as well as the output y, are expected to be a product of a distribution. The posterior expression is:

$$\text{Posterior} = \frac{(\text{Likelihood} * \text{Proir})}{\text{Normalization}}$$

- **Posterior:** The $P(B|y, X)$ conditional provides the responses as well as the predictive functions; the distributions among model variables are calculated depending upon certain information-driven likelihoods prior expert knowledge and belief.

As $n \to \infty$, the model parameters B converges to ordinary least squares linear regression.

- **Prior:** $P(B)$; we guess the model parameters over the predictor features based on expert knowledge and noninformative priors.
- **Likelihood:** $P(y, X|B)$; the probability overcomes the probability distributions whenever the amount of test data increases.

It is just the same as the Bayes' theorem, which stipulates

$$P\big(B|y, X\big) = \frac{P\big(y, X|B\big) P(B)}{P\big(y, X\big)}$$

$P(A)$ denotes the likelihood of event A happening, while $P(A|B)$ denotes the likelihood of A occurring if event B has already occurred. Since event B has already happened, $P(B)$ cannot be 0 ("Implementation of Bayesian Regression - GeeksforGeeks" 2020).

Benefits of Bayesian regression:

- Even though the data set is limited, it is very compelling.
- Particularly well-suited for online-based learning as contrasted with batch-based learning, since Bayesian Regression does not need data storage.
- The Bayesian methodology is proven to be computationally efficient. As a result, it can be used without any previous knowledge of the data set.

Pitfalls of Bayesian regression:

- The model's interpretation can take a long time.
- If our data set contains a lot of information, the Bayesian approach is ineffective, and the traditional frequentist approach is more successful.

Example:

Using a database containing academic performance with various variables, we will be constructing a model that predicts pupils' performance based on individual and educational features. In this case, the parameter's best estimate will be the average of every other model parameter from its paper trail. Likewise, any new point that is not in the test range can also be predicted ("Bayesian Linear Regression in Python: Using Machine Learning to Predict Student Grades Part 2 | by Will Koehrsen | Towards Data Science" 2018).

In the Bayesian regression approach, we use probability distributions instead of point estimates to perform linear regression; the response y is expected to be drawn from a probability distribution rather than being measured as a single value.

References

4 Logistic Regressions Examples to Help You Understand - Magoosh Data Science Blog (2018). Available at: https://magoosh.com/data-science/4-logistic-regressions-examples/

Bayesian Linear Regression in Python: Using Machine Learning to Predict Student Grades Part 2 | by Will Koehrsen | Towards Data Science (2018). Available at: https://towardsdatascience.com/bayesian-linear-regression-in-python-using-machine-learning-to-predict-student-grades-part-2-b72059a8ac7e (Accessed: 4 May 2021).

Implementation of Bayesian Regression - GeeksforGeeks (2020). Available at: https://www.geeksforgeeks.org/implementation-of-bayesian-regression/ (Accessed: 4 May 2021).

Lasso vs Ridge vs Elastic Net | ML - GeeksforGeeks (2020). Available at: https://www.geeksforgeeks.org/lasso-vs-ridge-vs-elastic-net-ml/ (Accessed: 3 May 2021).

Linear Regression in Machine learning - Javatpoint (2020). Available at: https://www.javatpoint.com/linear-regression-in-machine-learning

Logistic Regression in Machine Learning - Javatpoint (2020). Available at: https://www.javatpoint.com/logistic-regression-in-machine-learning

Machine learning - Why is logistic regression called regression? - Stack Overflow (2020). Available at: https://stackoverflow.com/questions/30499018/why-is-logistic-regression-called-regression

Machine learning Polynomial Regression - Javatpoint (2020). Available at: https://www.javatpoint.com/machine-learning-polynomial-regression

Regression with Regularization Techniques. | by Tarun Acharya | Towards Data Science (2019). Available at: https://towardsdatascience.com/regression-with-regularization-techniques-7bbc1a26d9ba (Accessed: 3 May 2021).

The Complete Guide to Linear Regression in Python | by Marco Peixeiro | Towards Data Science (2020). Available at: https://towardsdatascience.com/the-complete-guide-to-linear-regression-in-python-3d3f8f06bf8

What is LASSO Regression Definition, Examples and Techniques (2020) Great Learning Team. Available at: https://www.mygreatlearning.com/blog/understanding-of-lasso-regression/ (Accessed: 4 May 2021).

3

Methods to Predict the Performance Analysis of Various Machine Learning Algorithms

M. Saritha, M. Lavanya, and M. Narendra Reddy
Institute of Aeronautical Engineering

CONTENTS

3.1 Introduction

Measurement of performance is an essential part of the machine learning process. It is, however, a difficult process. Therefore, it must be carried out with caution, whether machine learning is used in internal medicine or other fields(Ramachandran et al., 2019). This chapter explains the problem and explores some of the most often used approaches to solving it. Performance modelling is a powerful equation that estimates and assesses the success of project management running on a certain computer or machine (typically in terms of execution time). Performance models can give vital information regarding constraints in efficiency. Machine learning, which is used to solve academic or commercial challenges, is becoming an increasingly important part of our lives (Awasthi et al., 2020).

DOI: 10.1201/9781003164265-3

To provide significant value to a company, machine learning algorithms must be able to make accurate predictions. While developing a model is a crucial, how the model generalises to new data is also an important factor to consider in any machine learning model. We really must know if it does work and, as a result, if we can trust the forecasts it makes. Is it possible that the algorithm is just remembering the input it is fed, without which it would be unable to make accurate predictions on unusual events or ones it has not seen before?

For legitimate reason, predictive models are becoming a trusted counsel to many organisations. These algorithms can "predict the future," so there are a variety of ways to choose from, and every industry should choose one that suits their needs (Kastrati & Imran, 2019). We are just referring to a regression analysis (continuous output) or a prediction model when we talk about predictive analytics (nominal or binary output). We utilise two types of algorithms in multiclass classification (depending on the type of output they produce):

- **Class output:** SVM and KNN methods provide a class result. In a binary classification issue, for example, the results would be either 0 or 1. Currently, though, algorithms have been developed that can transform these class results to probabilities.

- **Probability output:** Probability outputs are produced by techniques such as logistic regression, random forest, linear regression, Adaboost, and others. Establishing a cut-off likelihood is all it takes to convert probability results to class results.

While preparing data and building a machine learning algorithm are critical steps in the machine learning process, evaluating the learned performance of the models is as critical. Adaptive versus nonadaptive machine learning techniques are distinguished by how effectively the model generalises to new data (Rehman et al., 2021). We should really be able to enhance the overall prediction capability of our algorithm before rolling it out for production on unknown data by utilising multiple criteria for evaluation. When the relevant model is implemented on different datasets without sufficient analysis of the machine learning algorithm using different metrics and relying solely on accuracy, it might lead to problems and bad predictions. This occurs because, in situations like these, our models memorise rather than learn; as a result, they are unable to generalise successfully to new data (Narayanan et al., 2019).

In this chapter, we will go over the methods for determining how effectively machine learning generalises to new, previously unknown data. We will also learn how to use Python to construct standard model assessment metrics for regression and classification issues.

3.2 Analysis of Algorithms

Algorithm analysis is a crucial component of theory of computation, since it offers a conceptual estimate of how much resources and time an algorithm will need to solve a given task (*Design and Analysis of Algorithm – Tutorialspoint*, n.d.). Many programs are made to operate with inputs of any length. The assessment of the length of time and material resources necessary to perform an algorithm is known as algorithm analysis.

Typically, an algorithm's productivity or time slot is defined as a function linking the input duration to the series of phases (time complexity) or the memory space (space complexity) (*Analysis of Algorithms | Set 1 (Asymptotic Analysis) – GeeksforGeeks*, n.d.). In this section, we will learn about why it is important to analyse algorithms and how to pick the best algorithm for a given issue, because one computing problem can be handled by a variety of algorithms. By studying an approach for a perceived issue, we may begin to build image recognition, allowing this method to handle similar sorts of problems (*Analysis of Algorithms*, n.d.).

Algorithms are frequently extremely distinct from the others, even though their goals are the same. We know, for example, that a collection of integers may be ranked using several techniques. For much the same input, the number of samples done by one algorithm may differ from those of others. As a result, the temporal complexity of such methods may vary (*Analysis of Algorithms*, n.d.). At about the same time, we must determine how much RAM each method requires. Algorithm assessment is a strategy for examining an application's concern capabilities in terms of how lengthy and how bulling it is. However, the most important consideration in algorithm evaluation is the necessary time and performance.

3.3 Evaluation of Performance in Machine Learning Models

It is said that, "Evaluation is the key to making real progress in data mining (*Chapter 5. Credibility – Data Mining: Practical Machine Learning Tools and Techniques, 3rd Edition [Book]*, n.d.)." In this section, we will discuss the various types of evaluation methods and show some of the newer, lesser-known results in classifier evaluation. In this we'll also discuss how, over the last two decades, our definition of performance evaluation methods for machine-learned classifications has advanced in several ways (Flach, 2019).

Whether we use machine learning to solve educational and business challenges or not, it is becoming an important part of our lives (*Evaluating a Machine Learning Model.*, n.d.). In addition to adding significant value to business, machine learning techniques must be able to make an accurate prediction (*Different Metrics to Evaluate the Performance of a Machine Learning Model | by Swapnil Vishwakarma | Analytics Vidhya | Medium*, n.d.). Though building a classifier is a crucial stage, how the algorithm generalises to new data is also an important factor to consider in any machine learning workflow.

3.3.1 Methods for Model Evaluation

In this section we discuss various methods to calculate the performance of machine learning or deep learning models. We will also find out how to use one method in place of another. There are majorly nine different ways of calculating the performance of a model.

3.3.1.1 Confusion Matrix

A confusion matrix is a method used to summarise the performance of a particular system. The suggested model's performance is evaluated using an $N \times N$ matrix, where N is the number of class labels (*Simple Guide to Confusion Matrix Terminology*, n.d.). The matrix

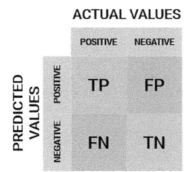

FIGURE 3.1
Binary classification matrix (*Confusion Matrix for Machine Learning*, n.d.).

determines the current goal values to the machine/deep learning model projections. This gives us a clear view of how well our classification model is performing and what kinds of mistakes it produces (*Confused About the Confusion Matrix? Learn All About It – Kambria*, n.d.). For example, let us consider a binary, i.e., 0 and 1 classification problem, which is demonstrated in Figure 3.1. It shows that the binary classification matrix has a 2×2 matrix with four values (TP, FP, FN, TN).

In the Figure 3.1, the class labels are positive and negative; columns represent the actual values of the class labels and rows represent the predicted values of the class labels. TP (True Positive) indicates that the predicted values and actual value are same, i.e., when both the values are predicted positive. FP (False Positive), also known as Type 1, indicates that the predicted values is negative but the actual value is positive. TN (True Negative) indicates that the actual values and predicted values are the same, i.e., when both the values are predicted negative. FN (False Negative), also known as Type 2 error, indicates that the predicted values is positive but the actual value is negative.

3.3.1.2 Accuracy

The percentage of accurate classification of the testing data is known as accuracy. It is simple to figure out: simply split the number of right guesses and the overall number of forecasts (*Machine Learning Model Accuracy | Model Accuracy Definition*, n.d.). The accuracy of a machine learning algorithm is a metric for determining whether an algorithm is better at finding patterns and correlations within observed variables based on the input or training data The faster a model can generalise to "unknown" data, the more accurate are its decisions and observations, and hence be the most value-adding (*Classification: Accuracy | Machine Learning Crash Course*, n.d.).

$$\text{Accuracy} = \frac{\text{Number of Correct Predictons (TP + TN)}}{\text{Total Number of Predictions made (TP + TN + FP + FN)}}$$

Formula 1: Calculating accuracy.

For example, let us consider 95% data in class X and 5% data in class Y of our training dataset. The model can easily get 95% accuracy just by predicting every training value in class X. Let us consider the same model tested on a testing set with 70% in class X and 30% in class Y, then the test accuracy will drop down to 70%. Classification accuracy is excellent, but it creates the impression that we have achieved high accuracy.

3.3.1.3 Precision

Precision is defined as the ratio between the TP and all the positives, i.e., TP and FP (*Precision vs Recall | Precision and Recall Machine Learning*, n.d.).

$$\text{Precision} = \frac{\text{True Positive (TP)}}{\text{True Positive(TP)} + \text{False Positive(FP)}}$$

Formula 2: Calculating precision.

For example, let us consider that from the data we obtained 46 TP and 37 FP, then from the above formula we get a precision of 0.5542. Precision also allows us to calculate the number of specific sample points (*How to Calculate Precision, Recall, and F-Measure for Imbalanced Classification*, n.d.).

3.3.1.4 Recall

Recall is a metric of how well the algorithm detects True Positives. Recall is the ratio between True Positive TP and the sum of True Positive TP and False Negative FN. Positive occurrences as a percentage of overall number of positive cases (*Classification: Precision and Recall | Machine Learning Crash Course*, n.d.). As a result, the denominator (TP+FN) equals the total number of positive cases in the collection.

$$\text{Recall} = \frac{\text{True Positive(TP)}}{\text{True Positive(TP)} + \text{False Negative(FN)}}$$

Formula 3: Calculating recall.

Recall is a metric that indicates how well our model can detect necessary data. It is also known as True Positive Rate or Sensitivity. For example, the recall value can be 0.83%

3.3.1.5 Specificity

This is the opposite of recall. Specificity is the ratio between the True NegativeTN and the sum of True NegariveTN and Fasle PositiveFP (*ML Metrics: Sensitivity vs. Specificity – DZone AI*, n.d.). As a result, the denominator (TN+FP) equals the total number of negative occurrences in the collection. It is comparable to recollection, except the emphasis is on unpleasant events. For example, how often healthy people were told they did not have cancer when they did not. It is a kind of test to check how distinct the categories are.

$$\text{Specificity} = \frac{\text{True Negative(TN)}}{\text{True Negative(TN)} + \text{False Positive(FN)}}$$

Formula 4: Calculating specificity.

3.3.1.6 F-score

F-score, which is also known as F1-score, is a performance evaluation measure for finding out the accuracy of a dataset (*F-Score Definition | DeepAI*, n.d.). F-score is basically used for the evaluation of the binary classification system, which is used to predict examples as being either "negative" or "positive." The F-score is a popular metric for assessing data mining techniques, such as search results, as well as a variety of machine learning algorithms, particularly in natural language processing (*Precision, Recall, and F Score Concepts in Details – Regenerative*, n.d.). It is possible to tweak the F-score such that accuracy takes precedence over recall, or vice versa. The F0.5-score and the F2-score, and even the normal F1-score, are common modified F-scores.

$$F1\text{-score} = \frac{2 * \text{precision} * \text{recall}}{\text{precision} + \text{recall}}$$

Formula 5: Calculating F-score.

   One disadvantage is that accuracy and recall are considered equal, which means that depending on the application, one may be more important than another, and F1-score may not be the best metric for it (Hand et al., 2021). As a result, using a weighted F1-score or looking at the PR or ROC curve can be beneficial.

3.3.1.7 ROC Curve

ROC (receiver operating characteristic) curves are frequently used to show in a graphical way the connection/trade-off between clinical sensitivity and specificity for every possible cut-off for a test or a combination of tests. For different threshold levels, a TPR versus FPR curve is shown (*Classification: ROC Curve and AUC | Machine Learning Crash Course*, n.d.). FPR rises in lockstep with TPR. In addition, the area under the ROC curve gives an idea about the benefit of using the test(s) in question. The ROC curve is created by comparing the TP against the FP for a classification algorithm for species that are widely distributed (*Understanding AUC – ROC Curve | by Sarang Narkhede | Towards Data Science*, n.d.). The projected likelihood of a data pertaining to the TP, for example, would have been the cut off in logistic regression. Generally, an AUC of 0.5 indicates no discrimination (i.e., the capacity to classify individuals with or without a psychiatric condition based on the result), 0.7–0.8 indicates good performance, 0.8–0.9 indicates great performance, and more than 0.9 indicates remarkable performance (*Understanding AUC – ROC Curve | by Sarang Narkhede | Towards Data Science*, n.d.).

$$\text{True Positive Rate(TPR)} = \text{Recall} = \frac{\text{True Positive(TP)}}{\text{True Positive(TP)} + \text{False Positive(FP)}}$$

$$\text{False Positive Rate(FPR)} = 1 - \text{Specificity} = \frac{\text{FP}}{\text{TN} + \text{FP}}$$

Formula 6: Calculating ROC.

3.4 Evaluation of Performance of Regression Model

In machine learning, model assessment is critical. It assists you in comprehending the performance of the models and makes it simple to display your model to others. There are several evaluation criteria available, and only a few of them seem to be acceptable for regression (*3 Best Metrics to Evaluate Regression Model? | by Songhao Wu | Towards Data Science*, n.d.). The multiple measures for the regression model, and the differences amongst them, will be discussed in Section 3.4. For example, "Hi, I developed a machine learning method to forecast Xyzzy," I tell my colleagues every time. "Nice, then what is the validity of your prediction models?" would be their initial reaction (*Evaluate the Performance of a Regression Model – Improve the Performance of a Machine Learning Model – OpenClassrooms*, n.d.). In contrast to classification, demonstrating performance in a regression model is a little more difficult. We will not be able to anticipate the exact amount, but we will be able to see how close our forecast is to the actual value.

There is a significant distinction in the approaches for assessing a predictive and classifying model. In regression, we deal with continuous data, and several approaches are used to try to figure out where the difference between the real and projected value is (*Evaluation Metrics for Your Regression Model – Analytics Vidhya*, n.d.). Whenever we try to evaluate a predictive model, however, we concentrate on the number of observations that are properly categorised. To properly assess a predictive model, we must additionally examine the pieces of data that were wrongly categorised. We often deal with two different classifier models: those that generate category output, such as KNN and SVM, where even the outcome is merely the class label, and those that generate probabilistic output, such as logistic regression, random forest, and others, where even the outcome is the likelihood of a dataset assigned to a single class, which we can convert using a cut-off value (*How to Evaluate Regression Models? | by Vimarsh Karbhari | Acing AI | Medium*, n.d.). In regression, there are three primary measures to evaluate models, which are briefly explained in Sections 3.4.1–3.4.3.

3.4.1 R Square or Adjusted R Square

R square is a metric for how well a model can explain variation in a predictor variable. It is named *R* Square since it is the square of the Correlation Coefficient (*R*) (*Regression Analysis: How Do I Interpret R-Squared and Assess the Goodness-of-Fit?*, n.d.). The sum of the squared error rate scaled by the entire sum of the square, which replaces the computed forecast with mean, yields *R* square. The value of *R* square ranges from 0 to 1, with a higher value indicating a better match between the forecast and the actual value.

$$R^2 = 1 - \frac{SS_{Regression}}{SS_{Total}} = 1 - \frac{\sum (y - y')^2}{\sum (y - y')^2}$$

Formula 7: Calculating *R* square (*3 Best Metrics to Evaluate Regression Model? | by Songhao Wu | Towards Data Science*, n.d.).

   R square is a useful metric for evaluating how much a model fits to given dependent variables. It does not, however, consider the problem of overfitting. Because the model is excessively intricate, if the regression model contains numerous independent variables, this may fit extremely well with the training data but perform badly on test dataset (*How to Interpret R-Squared in Regression Analysis – Statistics By Jim*, n.d.). Adjusted *R* square was created to punish the addition of extra independent variables to the model and modify the measure to avoid overfitting concerns. In Python we can calculate *R* square by using sklearn or statsmodel packages.

3.4.2 Mean Square Error or Root Mean Square Error

Since R square is a subjective concept of the model's ability to fit a regression model, mean square error (MSE) is an absolute standard of the fit's quality. MSE is computed by dividing the amount of data by the sum of the squares of error rate, which is real output minus expected output (*What Are Mean Squared Error and Root Mean Squared Error? – Technical Information Library*, n.d.). It provides an absolute number indicating how far your projected outcomes differ from the actual value. Although one single answer cannot provide many insights, it does provide an exact figure to contrast to these other results obtained and aid in the selection of the optimal regression model.

$$\text{MSE} = \frac{1}{N}\sum_{i=1}^{N}(y-y')^2$$

Formula 8: Calculating MSE (*3 Best Metrics to Evaluate Regression Model? | by Songhao Wu | Towards Data Science*, n.d.).

The square root of MSE is root mean square error (RMSE). It is more often used over MSE since the MSE number might sometimes be too large to compare readily (*What Are Mean Squared Error and Root Mean Squared Error? – Technical Information Library*, n.d.). Second, MSE is generated using the square of inaccuracy, thus taking the absolute value returns it to the same level of random variable and simplifies understanding. MSE and RMSE can be calculated from sklearn package in Python language (*MAE, MSE, RMSE, Coefficient of Determination, Adjusted R Squared — Which Metric Is Better? | by Akshita Chugh | Analytics Vidhya | Medium*, n.d.).

3.4.3 Mean Absolute Error

The mean absolute error (MAE) is comparable to the MSE. However, MAE takes the total of the actual number of errors rather than the sum of squares of error like MSE does (*MAE, MSE, RMSE, Coefficient of Determination, Adjusted R Squared — Which Metric Is Better? | by Akshita Chugh | Analytics Vidhya | Medium*, n.d.).

$$\text{MAE} = \frac{1}{N}\sum_{i=1}^{N}|y-y'|$$

Formula 9: Calculating MAE (*3 Best Metrics to Evaluate Regression Model? | by Songhao Wu | Towards Data Science*, n.d.).

MAE is a more straightforward depiction of the sum of all error terms than MSE or RMSE. MSE penalises high prediction mistakes by squaring them, but MAE handles all errors equally.

3.5 Examples

This section will help with understanding the coding of the performance evaluation measures discussed in Sections 3.3 and 3.4. Section 3.5.1 will explain the coding of performance evaluation of machine learning and Section 3.5.2 will explain the coding of performance measures of regression models.

3.5.1 Coding Example of Evaluation of Performance in Machine Learning Models

Let us consider the dataset that is easily available at the UCI repository: Heart Disease Dataset. Using the supplied collection of characteristics, we must determine if a patient has a cardiac condition or not. We will generate predictions using the simplest classifier – the KNN classification model – because this chapter is only about model assessment metrics.

Coding example in Python (Figures 3.2–3.5):

3.5.2 Coding Example of Evaluation of Performance of Regression Model

Let us consider an example of an incinerator dataset, which is available in the UCI repository or Kaggle. First, let us perform a regression model on the given dataset. In the below example the linear regression is done and based on that model the performance is evaluated.

Coding example in Python (Figure 3.6):

```
import numpy as np
from sklearn.model_selection import train_test_split
from sklearn.preprocessing import StandardScaler
from sklearn.neighbors import KNeighborsClassifier
from sklearn.metrics import confusion_matrix
from sklearn.metrics import classification_report
from sklearn.metrics import roc_curve
from sklearn.metrics import roc_auc_score
from sklearn.metrics import precision_recall_curve
from sklearn.metrics import auc
import matplotlib.pyplot as plt
import seaborn as sns
%matplotlib inline

import pandas as pd
data_df = pd.read_csv('heart.csv')
data_df.head()
```

	age	sex	cp	trestbps	chol	fbs	restecg	thalach	exang	oldpeak	slope	ca	thal	target
0	63	1	3	145	233	1	0	150	0	2.3	0	0	1	1
1	37	1	2	130	250	0	1	187	0	3.5	0	0	2	1
2	41	0	1	130	204	0	0	172	0	1.4	2	0	2	1
3	56	1	1	120	236	0	1	178	0	0.8	2	0	2	1
4	57	0	0	120	354	0	1	163	1	0.6	2	0	2	1

FIGURE 3.2
Importing all the required performance evaluation packages.

```
y = data_df["target"].values
x = data_df.drop(["target"], axis = 1)
#Scaling - mandatory for knn
from sklearn.preprocessing import StandardScaler
ss = StandardScaler()
x = ss.fit_transform(x)
#SPlitting into train and test
X_train, X_test, y_train, y_test = train_test_split(x, y, test_size = 0.3)
```

```
train_score = []
test_score = []
k_vals = []

for k in range(1, 21):
    k_vals.append(k)
    knn = KNeighborsClassifier(n_neighbors = k)
    knn.fit(X_train, y_train)

    tr_score = knn.score(X_train, y_train)
    train_score.append(tr_score)

    te_score = knn.score(X_test, y_test)
    test_score.append(te_score)
```

```
## score that comes from the testing set only
max_test_score = max(test_score)
test_scores_ind = [i for i, v in enumerate(test_score) if v == max_test_score]
print('Max test score {} and k = {}'.format(max_test_score * 100,
                                  list(map(lambda x: x + 1,
                                          test_scores_ind))))
```

```
Max test score 86.81318681318682 and k = [7]
```

FIGURE 3.3
Performing KNN and finding the test score.

```
#Setup a knn classifier with k neighbors
knn = KNeighborsClassifier(3)

knn.fit(X_train, y_train)
knn.score(X_test, y_test)
```

0.8241758241758241

```
y_pred = knn.predict(X_test)
confusion_matrix(y_test,y_pred)
pd.crosstab(y_test, y_pred, rownames = ['Actual'], colnames =['Predicted'], margins = True)
```

Predicted	0	1	All
Actual			
0	32	9	41
1	7	43	50
All	39	52	91

```
print(classification_report(y_test, y_pred))
```

	precision	recall	f1-score	support
0	0.82	0.78	0.80	41
1	0.83	0.86	0.84	50
accuracy			0.82	91
macro avg	0.82	0.82	0.82	91
weighted avg	0.82	0.82	0.82	91

FIGURE 3.4
Confusion matrix and classification report.

```
y_pred_proba = knn.predict_proba(X_test)[:,1]
fpr, tpr, thresholds = roc_curve(y_test, y_pred_proba)
```

```
roc_auc_score(y_test, y_pred_proba)
```

```
0.8982926829268293
```

```
precision, recall, thresholds = precision_recall_curve(y_test, y_pred_proba)

plt.figure(figsize = (10,8))
plt.plot([0, 1], [0.5, 0.5],'k--')
plt.plot(recall, precision, label = 'Knn')
plt.xlabel('recall')
plt.ylabel('precision')
plt.title('Knn(n_neighbors = 8) PRC curve')
plt.show()
```

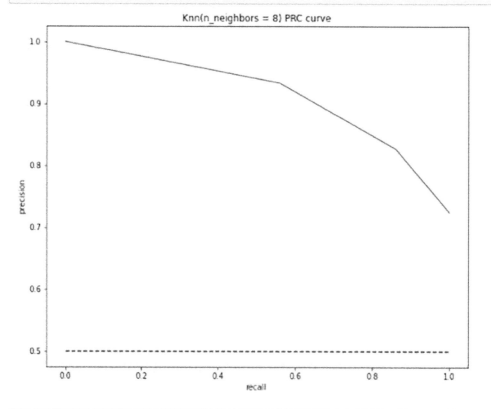

```
# calculate precision-recall AUC
auc_prc = auc(recall, precision)
print(auc_prc)
```

```
0.9139810479375697
```

FIGURE 3.5
Calculation of ROC, recall, and precision values.

```
y_pred_proba = knn.predict_proba(X_test)[:,1]
fpr, tpr, thresholds = roc_curve(y_test, y_pred_proba)
```

```
roc_auc_score(y_test, y_pred_proba)
```

```
#importing the dataset using pandas.
import pandas as pd

data = pd.read_excel ('Incinerator.xlsx')
#importing the dataset using pandas.
import pandas as pd

data = pd.read_excel ('Incinerator.xlsx')
```

```
import numpy as np
from sklearn.linear_model import LinearRegression
X = data['price'].values.reshape(-1,1)
y = data['age'].values.reshape(-1,1
```

```
from sklearn.model_selection import train_test_split
X_train, X_test, y_train, y_test = train_test_split(X, y, test_size=0.2, random_state=0)
```

```
regressor = LinearRegression()
```

```
regressor.fit(X,y)
```

```
LinearRegression()
```

```
y_pred = regressor.predict(X)
```

```
print(regressor.intercept_)
```

```
[42.04288224]
```

```
print(regressor.coef_)
```

```
[[-0.00025009]]
```

```
from sklearn import metrics
print('Mean Absolute Error:', metrics.mean_absolute_error(y, y_pred))
print('Mean Squared Error:', metrics.mean_squared_error(y, y_pred))
print('Root Mean Squared Error:', np.sqrt(metrics.mean_squared_error(y, y_pred)))
```

```
Mean Absolute Error: 18.15116629126008
Mean Squared Error: 940.7449008691028
Root Mean Squared Error: 30.67156502151631
```

```
from sklearn.metrics import r2_score
r2 = r2_score(y,y_pred)
print(r2)
```

```
0.11017989714144927
```

```
n=40
k=2
adj_r2_score = 1 - ((1-r2)*(n-1)/(n-k-1))
print(adj_r2_score)
```

```
0.06208151320314914
```

FIGURE 3.6
Performance evaluation in regression model.

References

3 Best metrics to evaluate Regression Model? | by Songhao Wu | Towards Data Science. (n.d.). Retrieved July 12, 2021, from https://towardsdatascience.com/what-are-the-best-metrics-to-evaluate-your-regression-model-418ca481755b

Analysis of Algorithms. (n.d.). Retrieved July 12, 2021, from https://aofa.cs.princeton.edu/10analysis/

Analysis of Algorithms | Set 1 (Asymptotic Analysis) – GeeksforGeeks. (n.d.). Retrieved July 12, 2021, from https://www.geeksforgeeks.org/analysis-of-algorithms-set-1-asymptotic-analysis/

Awasthi, K., Nanda, P., & Suma, K. V. (2020). Performance analysis of Machine Learning techniques for classification of stress levels using PPG signals. *Proceedings of CONECCT 2020–6th IEEE International Conference on Electronics, Computing and Communication Technologies.* https://doi.org/10.1109/CONECCT50063.2020.9198481

Chapter 5. Credibility - *Data Mining: Practical Machine Learning Tools and Techniques*, 3rd Edition [Book]. (n.d.). Retrieved May 24, 2021, from https://www.oreilly.com/library/view/data-mining-practical/9780123748560/xhtml/c0005.html

Classification: Accuracy | Machine Learning Crash Course. (n.d.). Retrieved July 12, 2021, from https://developers.google.com/machine-learning/crash-course/classification/accuracy

Classification: Precision and Recall | Machine Learning Crash Course. (n.d.). Retrieved July 12, 2021, from https://developers.google.com/machine-learning/crash-course/classification/precision-and-recall

Classification: ROC Curve and AUC | Machine Learning Crash Course. (n.d.). Retrieved July 12, 2021, from https://developers.google.com/machine-learning/crash-course/classification/roc-and-auc

Confused About The Confusion Matrix? Learn All About It – Kambria. (n.d.). Retrieved July 12, 2021, from https://kambria.io/blog/confused-about-the-confusion-matrix-learn-all-about-it/

Confusion Matrix for Machine Learning. (n.d.). Retrieved July 12, 2021, from https://www.analyticsvidhya.com/blog/2020/04/confusion-matrix-machine-learning/

Design and Analysis of Algorithm – Tutorialspoint. (n.d.). Retrieved July 12, 2021, from https://www.tutorialspoint.com/design_and_analysis_of_algorithms/analysis_of_algorithms.htm

Different metrics to evaluate the performance of a Machine Learning model | by Swapnil Vishwakarma | Analytics Vidhya | Medium. (n.d.). Retrieved July 12, 2021, from https://medium.com/analytics-vidhya/different-metrics-to-evaluate-the-performance-of-a-machine-learning-model-90acec9e8726

Evaluate the Performance of a Regression Model – Improve the Performance of a Machine Learning Model – OpenClassrooms. (n.d.). Retrieved July 12, 2021, from https://openclassrooms.com/en/courses/6401081-improve-the-performance-of-a-machine-learning-model/6519016-evaluate-the-performance-of-a-regression-model

Evaluating a Machine Learning Model. (n.d.). Retrieved July 12, 2021, from https://www.jeremyjordan.me/evaluating-a-machine-learning-model/

Evaluation Metrics for Your Regression Model – Analytics Vidhya. (n.d.). Retrieved July 12, 2021, from https://www.analyticsvidhya.com/blog/2021/05/know-the-best-evaluation-metrics-for-your-regression-model/

F-Score Definition | DeepAI. (n.d.). Retrieved July 12, 2021, from https://deepai.org/machine-learning-glossary-and-terms/f-score

Flach, P. (2019). Performance evaluation in machine learning: the good, the bad, the ugly, and the way forward. *Proceedings of the AAAI Conference on Artificial Intelligence*, 33(01), 9808–9814. https://doi.org/10.1609/aaai.v33i01.33019808

Hand, D. J., Christen, P., & Kirielle, N. (2021). F*: an interpretable transformation of the F-measure. *Machine Learning*, 110(3), 451–456. https://doi.org/10.1007/S10994–021–05964–1

How to Calculate Precision, Recall, and F-Measure for Imbalanced Classification. (n.d.). Retrieved July 12, 2021, from https://machinelearningmastery.com/precision-recall-and-f-measure-for-imbalanced-classification/

How to Evaluate Regression Models? | by Vimarsh Karbhari | Acing AI | Medium. (n.d.). Retrieved July 12, 2021, from https://medium.com/acing-ai/how-to-evaluate-regression-models-d183b4f5853d

How To Interpret R-squared in Regression Analysis – Statistics By Jim. (n.d.). Retrieved July 12, 2021, from https://statisticsbyjim.com/regression/interpret-r-squared-regression/

Kastrati, Z., & Imran, A. S. (2019). Performance analysis of machine learning classifiers on improved concept vector space models. *Future Generation Computer Systems*, 96, 552–562. https://doi.org/10.1016/J.FUTURE.2019.02.006

Machine Learning Model Accuracy | Model Accuracy Definition. (n.d.). Retrieved July 12, 2021, from https://www.datarobot.com/wiki/accuracy/

MAE, MSE, RMSE, Coefficient of Determination, Adjusted R Squared — Which Metric is Better? | by Akshita Chugh | Analytics Vidhya | Medium. (n.d.). Retrieved July 12, 2021, from https://medium.com/analytics-vidhya/mae-mse-rmse-coefficient-of-determination-adjusted-r-squared-which-metric-is-better-cd0326a5697e

ML Metrics: Sensitivity vs. Specificity – DZone AI. (n.d.). Retrieved July 12, 2021, from https://dzone.com/articles/ml-metrics-sensitivity-vs-specificity-difference

Narayanan, B. N., Ali, R., & Hardie, R. C. (2019). Performance analysis of machine learning and deep learning architectures for malaria detection on cell images. 11139, 111390W. https://doi.org/10.1117/12.2524681

Precision, Recall, and F Score Concepts in Details – Regenerative. (n.d.). Retrieved July 12, 2021, from https://regenerativetoday.com/precision-recall-and-f-score-concepts-in-details/

Precision vs Recall | Precision and Recall Machine Learning. (n.d.). Retrieved July 12, 2021, from https://www.analyticsvidhya.com/blog/2020/09/precision-recall-machine-learning/

Ramachandran, A., Ramesh, A., Pahwa, P., Atreyaa, A. P., Murari, S., & Anupama, K. R. (2019). Performance analysis of machine learning algorithms for fall detection. *2019 IEEE International Conference on E-Health Networking, Application and Services, HealthCom* 2019. https://doi.org/10.1109/HEALTHCOM46333.2019.9009442

Regression Analysis: How Do I Interpret R-Squared and Assess the Goodness-of-Fit? (n.d.). Retrieved July 12, 2021, from https://blog.minitab.com/en/adventures-in-statistics-2/regression-analysis-how-do-i-interpret-r-squared-and-assess-the-goodness-of-fit

Rehman, H. A. U., Lin, C.-Y., Mushtaq, Z., & Su, S.-F. (2021). Performance analysis of machine learning algorithms for thyroid disease. *Arabian Journal for Science and Engineering* 2021, 1–13. https://doi.org/10.1007/S13369–020–05206–X

Simple Guide to Confusion Matrix Terminology. (n.d.). Retrieved July 12, 2021, from https://www.dataschool.io/simple-guide-to-confusion-matrix-terminology/

Understanding AUC – ROC Curve | by Sarang Narkhede | Towards Data Science. (n.d.). Retrieved July 12, 2021, from https://towardsdatascience.com/understanding-auc-roc-curve-68b2303cc9c5

What are Mean Squared Error and Root Mean Squared Error? – Technical Information Library. (n.d.). Retrieved July 12, 2021, from https://www.vernier.com/til/1014

4

A Viewpoint on Belief Networks and Their Applications

G S Sivakumar, P Suneetha, and V Sailaja
Pragati Engineering College

Pokala Pranay Kumar
Woxsen University

CONTENTS

4.1 Introduction

Prediction models can be used to analyze and predict and create connections between elements. Completely conditioned systems, for instance, might necessitate a large quantity of data to encompass every conceivable scenario, and chances may be difficult to estimate in practice. Streamlining conditions like the independence between every pair of all probability distributions, as with naive Bayes, could be useful; however, it is a radical move.

Constructing a framework that maintains established conditional dependence among probability distributions and is conditionally independent in other circumstances is another option. Bayesian networks are a statistical hypothesis that expresses recognized conditional dependence inside a weighted graph having line segments. The conditional independencies within the framework are defined by all transitional forms. As a result, Bayesian networks are a powerful way of visualizing a property's prediction algorithm, reviewing all of the probability distribution interactions, and reasoning regarding causation possibilities for possibilities based on existing information [1].

A belief network is indeed a directional description of contingent dependency between a collection of explanatory variables. The directional cues are taken into consideration inside the exact description of independence between every pair inside a belief network. Several areas could be represented succinctly using the concept of conditional probability. The notion is that with an unknown parameter X, a limited collection of elements can arise that directly affect the value of the dependent variable, such that X is uncorrelated to many other values when the directly impacted elements have quantities. The Markov blanket is indeed a collection of spatially affecting variables. In a belief network, this specificity is leveraged.

DOI: 10.1201/9781003164265-4

We begin by defining a belief network using a collection of arbitrary parameters that reflect every one of the model's attributes. Assume that the parameters are $X_1, ..., X_n$. After that, choose a form filled with elements such as $X_1, ..., X_n$.

$$P(X_1 = v_1 \wedge X_2 = v_2 \wedge ... \wedge X_n = v_n) = \prod_{i=1}^{n} P(X_i = v_i | X_1 = v_1 \wedge ... \wedge X_{i-1} = v_{i-1}).$$

Parse conjunction down to conditional probabilities. In terms of random variables and probability distributions.

$$P(X_1, X_2, ..., X_n) = \prod_{i=1}^{n} P(X_i | X_1, ..., X_{i-1}).$$

Specify the randomized variable's origins. X_i is defined as a minimum subset of X_i's precursors in the whole arrangement given all the remaining X_i precursors are conditionally independent upon X_i provided parenting (X_i).

$$\text{i.e, parents}(X_i) \subseteq \{X_1, ..., X_{i-1}\} \text{ such that}$$

$$P(X_i | X_{i-1} ... X_1) = P(X_i | \text{parents}(X_i)).$$

If there are multiple fundamental sets, any of them would be picked as the parents. Just because some of the antecedents are probabilistic variables of someone else, there would be more than one smallest collection.

$$P(X_1, X_2, ..., X_n) = \prod_{i=1}^{n} P(X_i | \text{parents}(X_i))$$

The joint probability distribution is defined as the probability across all parameters, $P(X_1, X_2, ... X_n)$. The belief network is a decomposition of a joint probability distribution in which conditional probabilities are repeated together just to produce elements.

A belief network, also known as a Bayesian network, is a plane curve with probability distributions as a node. Every component of parents(X_i) has an arc that leads to X_i. A collection of conditional probability distributions – meaning conditional probability within every parameter knowing its parents – are connected with the belief network [2] (Figure 4.1).

In definition, a belief network is monoclinic. The ranking is decided by how well the chained principle understands what the conjunction entails. Just predecessors can be the parents of variables. Various decompositions can lead to various belief networks.

Example 1. Let's say we need to use the diagnostics assistance to determine if a building is now on flames relying on noise data sources and potentially contradicting descriptions of what is happening there. The operator gets a message asking if everybody is entering the hotel. Assume the reporting sensor is loud: it records departing but not evacuation (a false positive), but it does not indicate all the departure (a false negative). Assume that the sound of a smoke detector causes people to leave, but this is not a predictable relation. The security system could have been harmed by manipulation, a flame, or thick smoke from the structure as a result of the incident. The variables we can use are:

- Whenever the sensor is tampered with, it is also considered tampering.
- Whenever there is a fire, it is real.

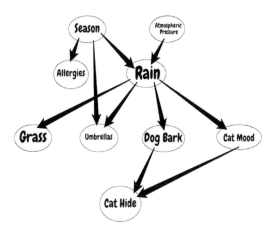

FIGURE 4.1
Example of belief networks. (Ref. [3])

- Whenever the alarm goes off, it is genuine.
- Whenever there is smoke, it is accurate.
- Leaving is true if many people are leaving the building at once.
- If somebody reports that individuals are departing, the claim is correct. When there is no record of departure, the claim is untrue.

The factor of change According to sensor reports, companies are fleeing. This data is untrustworthy since the individual delivering the information could be pranking or no one who could have supplied the information paid close attention. That parameter was included to accommodate for sensor information that was not entirely dependable. The agents are aware of the sensor's findings, although they possess shaky confirmation of humans exiting in the structure (Figure 4.2). The following are the independencies:

- A manipulation is an independent event of a Fire.
- Either flames and interference can raise an alarm. That is, with that parameter arrangement, we make no promises on how Warning is dependent on antecedents.
- Smoke is independent of events of Tampering and Warning depending on whether or not there would be a Fire.
- Leaving is solely dependent on the alert, not so much on burning, manipulation, or smoking. That is, assuming Alarm, Departure seems to be independent of other factors.
- Just Leaving has a significant effect on the assessment.

Belief networks are sometimes referred to as causal networks, and they are said to provide an excellent depiction of causality. The causal model predicts the outcome of operations, as you may recall. Assume you are considering causal modeling of a subject in which the area is composed of a set of explanatory variables. If there is a clear causative relationship between X_1 and X_2, add the arcs from X_1 to X_2. You would think that now the conceptual framework should follow the belief network's estimation methods. As a result, the belief network's findings would always become correct. You would likewise anticipate

FIGURE 4.2
Network design of the report. (Ref. [2])

an acyclic graph because it don't need anything to potentially cause oneself. When we assume that perhaps the random variables indicate specific occurrences instead of types of occurrences, the hypothesis makes sense. Take the following causative sequence: "feeling worried" leads to "working incompetently," which leads to "becoming frustrated." To interrupt this loop, one may express "feeling worried" in different phases as numerous different parameters referring to varying moments. Being anxious in the previous case drives one to perform poorly in the present, which leads to stronger distress. The parameters must adhere to the principle of clarity to have a good interpretation. These parameters are not to be confused with types of occurrences. The belief network has nothing to do with causation and can reflect non-causal independence, but it appears to be particularly well-suited to describing causality. When you include arcs that represent local causation, you have a smaller belief network.

4.2 Belief Networks Designing

To build belief networks one has to think about the following questions, which help them solve problems using belief networks:

1. What variables to consider.
2. Whatever the agent might notice throughout the environment. The entity has to be capable in learning, depends on most of its findings and every characteristic detected ought to be a parameter.
3. Considering the facts, what data would the agent want to find the value with?
4. What might the weights of such parameters be? These entails deciding mostly on the degree of complexity with which the agent must reason through ability to answer same kinds of questions to be asked.
5. What is the relationship between a variable's distribution and the elements that impact it directly (its parents)? The conditional probability distributions are used to describe concept.

Example: To identify robbery, Harry added a new security alarm in his house. The sensor is not only sensitive enough to detect a robbery, but it can also identify mild earthquakes. Harry has two next-door neighbors, David and Sophia, who have already agreed to notify Harry while he is at work if they hear an alert. Whenever David hears the alert, he typically calls Harry, but not always. However, at one point, he is bothered by the ringing phone and by phone calls. Sophia, on the other hand, enjoys listening to loud music and occasionally ignores the alarms. We would want to evaluate the probability of a robbery alarm in this case.

Problem: Compute the likelihood that perhaps the alert went off even though there was no robbery or disaster, yet David and Sophia simultaneously contacted Harry.

Solution: Again, the Bayesian network is employed to solve the previously described challenges. The communication network reveals that burglary and earthquakes have become the parent nodes of both the sensor, impacting the likelihood included its immediate warning, but David and Sophia's contacts were mostly reliant on the possibility of the persistent beeping out.

Our hypotheses, according to the system, don't really power of the united the robbery, need not detect the tiny earthquakes, and therefore do not communicate while contacting. Conditional probabilities tables, or CPTs, are used to represent the conditional distributions by each nodes.

Since all of the values throughout the database indicate a substantially faster of instances for both the parameter, every rows throughout the CPT should add to one. Two thousand probabilities are included in a Boolean variable having k Boolean parents using CPT. As a result, whether there are two people, CPT can have four parameter estimates (Figure 4.3).

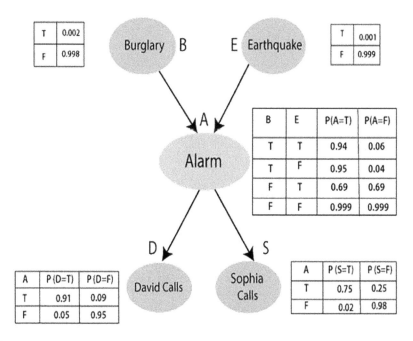

FIGURE 4.3
Probability of all the factors truth table. (Ref. [4])

List of all events occurring in this network:

- Burglary (*B*)
- Earthquake (*E*)
- Alarm (*A*)
- David calls (*D*)
- Sophia calls (*S*)

$$P[D, S, A, B, E]$$

$$= P\left[D \mid S, A, B, E\right] . P[S, A, B, E]$$

$$= P[D \mid S, A, B, E] . P[S \mid A, B, E] . P[A, B, E]$$

$$= P[D \mid A] . P\left[S \mid A, B, E\right] . P\left[A, B, E\right]$$

$$= P[D \mid A] . P[S \mid A] . P[A \mid B, E] . P[B, E]$$

$$= P\left[D \mid A\right] . P\left[S \mid A\right] . P\left[A \mid B, E\right] . P[B \mid E] . P[E]$$

4.3 Applications of Belief Networks

1. For the generation and recognition of objects, deep belief networks are being deployed [5].
2. Video sequences and motion-capture data [6].
3. Deep belief networks provide nonlinear dimensionality reduction and therefore can acquire compact binary digits, which enable extremely quick recovery of texts or pictures if indeed the number of vehicles inside the edge network is modest [7].

References

1. Brownlee J. A gentle introduction to Bayesian belief networks [Internet]. *Machine Learning Mastery.* 2019 [cited 2021 Jul 10]. Available from: https://machinelearningmastery.com/introduction-to-bayesian-belief-networks/
2. Poole D, Mackworth A. *Artificial Intelligence - foundations of computational agents --6.3 Belief Networks [Internet].* Cambridge University Press. 2017 [cited 2021 Jul 10]. Available from: https://artint.info/html/ArtInt_148.html
3. Güney A. Introduction to Bayesian Belief Networks [Internet]. *Towards Datascience.* 2019 [cited 2021 Jul 10]. Available from: https://towardsdatascience.com/introduction-to-bayesian-belief-networks-c012e3f59f1b

4. Bayesian belief network in artificial intelligenc [Internet]. *JavaTpoint*. [cited 2021 Jul 12]. Available from: https://www.javatpoint.com/bayesian-belief-network-in-artificial-intelligence
5. Hinton GE, Osindero O, Teh YW. A fast learning algorithm for deep belief nets. *Neural Comput [Internet]*. 2006 [cited 2021 Jul 12];18(7):1527–54. Available from: https://pubmed.ncbi.nlm.nih.gov/16764513/
6. Sutskever I, Hinton G. Learning multilevel distributed representations for high-dimensional sequences [Internet]. *PMLR*; 2007 [cited 2021 Jul 12]; 548–55. Available from: http://proceedings.mlr.press/v2/sutskever07a.html
7. Hinton GE, Salakhutdinov RR. Reducing the dimensionality of data with neural networks. *Science* (80-) [Internet]. 2006 Jul 28 [cited 2021 Jul 12];313(5786):504–7. Available from: https://science.sciencemag.org/content/313/5786/504/

5

Reinforcement Learning Using Bayesian Algorithms with Applications

H. Raghupathi and G Ravi
Visvesvaraya College of Engineering and Technology

Rajan Maduri
Vignan Institute of Science and Technology

CONTENTS

5.1 Introduction

Reinforcement learning is the process of teaching machine learning models to make a series of judgments. In an uncertain, possibly complicated environment, the agent learns to attain a goal. Machine intelligence meets a play circumstance in RL. To find a solution to the issue, the software needs different techniques. Machine learning is given either rewards or punishments and for its actions that it takes in order to get someone to accomplish whatever the developer desires. For each correct step towards the goal, the agent will get a reward.

Despite the fact that the designer sets the reward and punishment in the competition's rules, he didn't give any hints or ideas to handle this problem. Beginning with totally random attempts and progressing toward complex strategies and super abilities, it is indeed down to a model to find out how to perform the assignment in order to maximize the reward. RL is arguably the most popular technique to suggest mechanical inventiveness by leveraging the power of searching, with numerous repetitions. Unlike humans, artificial intelligence may gather experience through hundreds of thousands of perceptron's in a network when a reinforcement training RL algorithm is run on a powerful computational system [1].

5.1.1 Applications of Reinforcement Learning

The AWS DeepRacer is a driverless racing automobile that was created to put RL to the trial on a real-world racetrack. This controls the accelerator and orientation using an RL

DOI: 10.1201/9781003164265-5

model, but also webcams that visualize the airport. Wayve.ai have effectively used RL to teach an automobile to operate in just one day. To handle the lanes tracking problem, researchers developed a deep reinforcement learning algorithm. They used a deep network containing four convolutional layers, including three reasonably obtainable layers. The graphic throughout the foreground depicts the driver's point of view.

Training robots were used to execute many jobs in industrial reinforcement. Other than being highly effective than humans, such machines are capable of performing activities that could be harmful to people.

Deepmind's usage of AI agents to help cool Google Data Centers is indeed a great example. It resulted in a 40% decrease in amount of energy used for cooling. Artificial intelligence currently controls these facilities completely without human interaction.

Forecasting potential sales and equity markets are both done with supervised time series analysis. Such algorithms, on the other hand, do not select what to do with a given company's stock. Here is where RL comes in. The RL agent may select to either retain, purchase, or offer work. To ensure the RL algorithm is working efficiently, research methodology measures are used.

Unlike earlier techniques that required researchers to evaluate each and every choice, this system guarantees uniform performance throughout the procedure. IBM, for instance, has created a comprehensive RL-based platform that could execute monetary operations. Each business's proposed transaction losses or revenue is used to calculate the objective functions.

Individuals in medicine could benefit from strategies established through RL systems. Despite prior knowledge of both the physical equation of biological processes, RL can develop the best strategies based on historical encounters. It renders RL better suited for medicine than other technologies.

Because consumer tastes vary regularly, offering content to people based on recommendations and what they love can lead to information overload. The RL algorithm may monitor an individual writer's returning actions using reinforcement learning. Acquiring information attributes, readers attributes, contextual attributes, and user information attributes is essential to build a good solution. Contents, headlines, and sources are a few examples of media elements. A viewer's interaction only with the material, such as clicking and sharing, is referred to as consumer aspects. Media elements including timeliness and originality of the content are examples of background elements. Following this, compensation was determined on the basis of most of the users' actions.

The capacity to implement at an individual level is critical throughout marketing, because selecting the correct objectives ensures a good business model. Real-time bargaining in multi-agent reinforcement learning A clustered approach can manage a wide range of ads, with tactical negotiating agents allocated to each cluster.. A distribution coordinate multi-agent bidding is suggested that reconciles the exchange between competitiveness and collaboration between marketers.

Deep learning as well as reinforcement learning could be used to teach machines to handle any variety of tasks, such as those that aren't visible throughout learning. This could be employed, for instance, in a manufacturing line to produce things. This is achieved by utilizing large-scale distributed optimizing with a deep Q-Learning variation known as QT-Opt. Because QT-Opt supports complex search areas, it is well suited for automation challenges. A model is trained offline before being deployed and fine-tuned on a real robotic machines [2] (Figure 5.1).

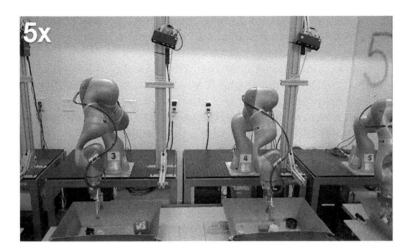

FIGURE 5.1
QT-Opt scalable deep reinforcement learning for vision-based robotic manipulations. (Ref. [2])

5.2 Bayesian Reinforcement Learning

The earliest kind of RL was Bayesian RL. Many information systems scientists investigated the topic of regulating Markov chains with variable chances as early as the 1950s and 1960s.

5.3 Model-Free Reinforcement Learning

The model-free RL approach uses sampling paths collected through immediate experience and interaction rather than learning a representation of the proposed.

Model-free approaches were typically simpler to integrate because they do not involve a database schema to describe a concept, or perhaps the algorithm constantly maintains it.

5.4 Value-Function-Based Algorithms

Value-function-based RL approaches look for the best values (action-value) combination inside the universe of the cost function, subsequently utilizing this to determine the right approach. We look at two Bayesian value-function-dependent RL algorithms in this segment:

- Bayesian Q learning (BQL)
- Gaussian process

The initial approach is optimized for domain-containing discrete-continuous state fields, whereas the other technique was designed for areas containing continuous movement of information.

We choose actions to undertake only on the basis of local Q-value information, as is the case with undirected exploration strategies. We may, however, make better informed judgments by retaining and propagating distributions across the Q-values rather than point estimations. Let $D(s, a)$ signify that the aggregate with discount rate eventually reached because once intervention a is conducted throughout states as well as optimized legislation is then implemented. $E[D(s,a)] = Q(s,a)$ is the standard Q-expectation parameter for this parameter.

We keep the hyper-parameters including its distributions across each D in BQL rather than the Q-values like in conventional Q-learning (s, a). As a result, BQL will only be used for Markov Decision Process (MDPs) having limited continuous action space inside their initial form. After executing a in s and monitoring r as well as s for each sampling interval, the distributions so over D's are defined as follows:

$$P\big(D(s,a)\|r,s'\big) ===\propto \int_d P\big(D(s,a)\big)P\big(\|r+\gamma d \mid D(s,a)\big)P\big(D(s',a')=d\big)$$

$$\propto \int_d P\big(D(s,a)\big)P\big(\|r+\gamma d \mid D(s,a)\big)P\big(D(s',a')=d\big)$$

The increase in worth should indeed be put into consideration if research leads to a policy decision. Value of Perfect Information (VPI) is typically calculated as just a potential gain because the individual does not know what will happen within each activity beforehand.

$$\text{VPI}(s,a) = \int_{-\infty}^{\infty} dx \; \text{Gain}(s,a)(x)P\big(Q(Sa,a)=x\big)$$

Here, overall gain is the gain produced by knowing the precise Q value of the activity being performed.

$$\text{Gains}_{s,a}\big(q_{s,a}\big) = \begin{bmatrix} q_{s,a} - E\big[Q_{(s,a1)}\big] \text{ if } a \ne a1 \quad \text{and} \quad q_{s,a} > E\big[Q_{(s,a1)}\big] \\ E\big[Q_{(s,a2)}\big] - q_{s,a} \text{ if } a = a1 \quad \text{and} \quad q_{s,a} < E\big[Q_{(s,a2)}\big] \\ 0 \quad \text{Otherwise} \end{bmatrix}$$

There are two options: a has a greater Q than operation a1 with the strongest correlation Q-value or a1 with the strongest correlation Q-value has a lesser Q-value then activity a2 with the second best projected Q-value.

When accessing services, we directly compute a description including its environment dynamics through model-based RL. In model-based Bayesian RL, you begin with some previous knowledge about the MDP model's uncertain variables. The hypothesis then gets updated to ensure the observed values whenever a manifestation of such a population quantity is noticed when communicating using that environment. Model-based methods are more computationally efficient than model-free techniques; however, they typically provide for even more intuitive integration of the existing knowledge about the environment into the training procedure.

Although Bayesian RL was probably the very first type of RL addressed by the Operational Scientific community throughout the 1960s, the Machine Learning community's current spike of enthusiasm has led to so many breakthroughs, which are presented in this section. The advantages of preserving precise distributions and having control over quantity for interests account for a large portion of such a motivation. When a distribution that characterizes the uncertainties surrounding different elements of the network, cost function, or gradient is utilized, the exploratory dichotomy could be automatically optimized. While optimizing a strategy, risk concepts could also be considered. We gave an assessment of the history of industry in using Bayesian approaches with reinforcement learning for just a single individual over predicted result settings throughout this chapter. In reinforcement learning domains and multi-agent technologies, Bayesian approaches were likewise successfully applied [3].

References

1. Osiński B, Budek K. What is reinforcement learning? The complete guide - deepsense.ai [Internet]. [cited 2021 Jul 13]. Available from: https://deepsense.ai/what-is-reinforcement-learning-the-complete-guide/
2. Mwiti D. 10 Real-life applications of reinforcement learning - neptune.ai [Internet]. Neptune.ai. [cited 2021 Jul 13]. Available from: https://neptune.ai/blog/reinforcement-learning-applications
3. Bittner LR. Bellman, adaptive control processes. A guided tour. XVI+255 S. Princeton, N. J., 1961. Princeton University Press. Preis geb. $ 6.50. ZAMM - *J Appl Math Mech / Zeitschrift für Angew Math und Mech [Internet]*. 1962 Jan 1 [cited 2021 Jul 13];42(7–8):364–5. Available from: https://onlinelibrary.wiley.com/doi/full/10.1002/zamm.19620420718

6

Alerting System for Gas Leakage in Pipelines

Nilesh Deotale, Pragya Chandra, Prathamesh Dherange, Pratiksha Repaswal, and Saibaba V More

G.H. Raisoni Institute of Engineering and Technology

CONTENTS

6.1 Introduction

Gas leakage is one of the most serious accidents that can occur in industrial establishments. This can result in a variety of losses, fatalities, and major equipment damage. To protect against gas leakage in our environment, sophisticated emergency management systems must be put in place. In this regard, we propose a leak detection system that collects sensor data with effective parameters in real time and detects leakage in the pipeline using

DOI: 10.1201/9781003164265-6

machine learning and ensures that the system is completely automated. A pipeline system is the cheapest and safest way to transport gases, but pipelines are being constructed at an increasing pace, increasing the risk of gas leaks. Gas leak is a major problem because it can result in a large number of human deaths, as well as cause economic and environmental damage. Gas leaks can occur for a variety of reasons, including pipe corrosion, installation defects, pipeline quality, and operating conditions. In the 1984 Bhopal gas tragedy, nearly 3,787 people died and approximately 574,366 were injured. It was one of the worst industrial disasters in history. To avoid such mishaps, it is crucial to build a system that can detect and notify gas leaks in pipelines. There are several approaches available to avoid such risks; however, the aquatic approach for detecting leakage is one step ahead of other techniques [1]. The author proposed in this paper a method that uses the action of sound in pipelines to locate and detect leaks. This method looks for trends in the sound signal detected by the sensor and classifies the result into three groups based on the signal data system: leakage, no leakage, or leakage (external disturbance). There are numerous instruments or sensors on the market that detect gas leakage in pipelines, but there is still room for improvement, and research is ongoing to make the instruments more reliable and user friendly. In the world of artificial intelligence and machine learning, there is a wide range of applications where these techniques can be used to track, avoid, and prevent leakage in pipes [2,3]. In this chapter, we proposed a model that detects gas leakage in pipelines using sensor data and a machine learning approach, and we compare the results of our proposed model to other recognized works in this area.

6.2 Previous Work

6.2.1 Machine Learning and Acoustic Method Applied to Leak Detection and Location in Low-Pressure Gas Pipelines

Rodolfo Pinheiro da Cruz · Flávio Vasconcelos da Silva Ana Maria Frattini Fileti

In this paper the acoustic method is used along with machine learning techniques for leakage detection under low pressure. The method is based on the fact that the sound behavior of pipelines changes when a leak occurs. This behavior is monitored and used to identify and locate leakages. There were three possible outcomes during the monitoring of the pipelines: leak, no leak, and possible leak.

The acoustic method struggles to detect small leakages, and false alarms can occur when there are external interferences; furthermore, the lower the pressure in the pipeline, such as in gas distribution networks, the more serious the difficulties. To address these obstacles, an approach that combines the use of microphones and machine learning algorithms is proposed in the present work.

The algorithms employed in this paper were logistic regression, K-nearest neighbors (KNN), support vector machines with linear kernel (SVM-L), support vector machine with radial basis kernel (SVM-RBF), random forests (RF), adaptive boosting (AdaBoost), and extreme gradient boosting (Xgboost).

When all the tests were concluded, 1,800,000 observations of sound amplitudes in the time domain were collected from each of the four microphones for every one of the 14 experiments. Fast Fourier transform (FFT) was used to convert the data into the frequency domain. The random forest algorithm was employed to find the most important features present in the data. The importance of a feature is measured by the decrease in impurity that it produces. All models were built in the Scikit-learn library. There were methods designed for classification and regression. The classification algorithms were fed with data from all the 14 tests. The regression models were supplied only with data from the one leakage experiment. Their task was to predict the location of the leakages. The dataset for classification consisted of 1,680 samples and for regression 840 samples.

The data were split, 80% of the observations for training and 20% for evaluation of the models, and then scaled. Two other metrics were used to evaluate the classification models *precision* and *recall*. In this context, high precisions indicated low rates of false alarms and high recalls a small percentage of leakages not being detected. The classification algorithms were trained with the three no leakage experiments and a set of the one leakage experiment. The trained model was then asked to make predictions for the unseen data from the one leakage experiment. A similar procedure was carried out for the regression algorithms, but in this case, the no leakage experiments were not used during the training procedure, only six of the seven one leakage and the value predicted by the model were compared with the actual value of the experiment not used on training. Three metrics were used to measure the performance of the classification models: accuracy, precision, and recall. The results also recommend the use of random forest classifiers to detect leakages, which had the best performance in all tests that were conducted. Regarding the performance of the models, the AdaBoost algorithm had the best one, which corresponds to the lowest root mean square error (RMSE) [4]. The small value of the RMSE obtained with the AdaBoost model means that the predictions made by it were very close to the actual positions of the orifices. The highest location error was 0.42%, and the average for the seven positions was 0.1%. The location errors were higher when the models were asked to make predictions to data from experiments that were not in the training procedure.

The experimental results showed that the method was efficiently detecting leakages; 99.6% of the leakage events were correctly classified. Leaks came from orifices as small as 0.5 mm in diameter. As important as the capacity of identifying leakages was the low rate of false alarms achieved: only 0.3% of the no leakage events were incorrectly classified as leakages.

6.2.2 Detection and Online Prediction of Leak Magnitude in a Gas Pipeline Using an Acoustic Method and Neural Network Data Processing

R. B. Santos, E. O. de Sousa, F. V. da Silva, S. L. da Cruz and A. M. F. Fileti*

In this paper, the characteristics of the noise generated by gas leakage in a pressure vessel pipeline system were analyzed. A domestic-type LPG vessel was used as the pressure vessel. In this work the pipeline operated under nominal pressures of 4 and 6 kgf/cm^2, with the gas leaking through orifices of 1, 1.5, 2, 2.5, and 3 mm in diameter [5] The monitoring of leaks was performed through a microphone, installed in the pressure vessel, and connected to a microcomputer. The first phase of this work consisted of collecting the leak detection data by using the acoustic method, generating data for training the neural model. With the experimental data coming from the acoustic system (microphone and signal conditioning circuit), noises in different frequencies were obtained. These noises were processed in the filter bank, resulting in three voltage signals with frequency bands of 1, 5 and 9 kHz. The dynamics of these noises in time was used as input to the neural network model in order to determine the magnitude of the leak. Network training was carried out with data obtained with and without leak occurrence. The number of data patterns used for training was 23,166 and 23,504 for the pressures of 4 and 6 kgf/cm², respectively.

The artificial neural network training program used in this study was implemented in MATLAB software. The method chosen for training the neural network was the Levenberg-Marquardt with Bayesian regularization ("trainbr" function in MATLAB) to avoid overfitting of the neural model.

At the time that the leak was provoked, the variation in the amplitude of the noise allowed leakage to be readily detected. The amplitude of the noise increased abruptly and remained relatively constant, reaching a new steady state. Two neural models have been developed in order to determine the occurrence and the magnitude of the leaks occurring at the inlet end of the pipeline, under nominal pressures of 4 and 6 kgf/cm². It is observed that, for all cases, the neural model was able to detect the leak, but there was a delay of approximately 5.44 s for the case of a leak occurring through the 1 mm diameter orifice.

The neural model showed 100% accuracy in the detection of leaks in the online tests with a nominal pressure of 6 kgf/cm².

6.2.3 Experimental Study on Leak Detection and Location for Gas Pipeline Based on Acoustic Method

Lingya Meng, Li Yuxing*, Wang Wuchang, Fu Juntao

When a leak or rupture occurs, the pressure balance of the pipeline is interrupted and gas (or fluids) leak from the pipeline and acoustic waves are generated by the friction with the pipe wall. Here, wavelet transform and Fourier transform are applied to signal de-noising, and signal to noise ratio (SNR) and value of root mean square (RMS) are considered as the criteria to judge the effectiveness of the de-noising. And the effect of de-noising becomes better with larger SNR and smaller RMS. Once a leak occurs, acoustic signals are generated at the leak point, which are detected by the acoustic sensors at both upstream and downstream. When a leak occurs, the upstream acoustic signal is similar to the downstream signal but with a time difference [6]. Therefore cross-correlation analysis is conducted by moving the acoustic signal from one end and comparing with the signal on the other end. There must be a maximum point; once that is found, the time difference of the acoustic waves between upstream and downstream can be calculated. By calculating the maximum value of the cross-correlation function, the leak location can be determined.

As further processing of acoustic signals is done, time-frequency analysis is used to analyze the characteristics of the frequency domain at Adjacent intervals. From the time-frequency analysis of acoustic signals, the amplitude and power spectral density of signals can be used as the characteristics for leak detection. The accumulated value difference (sumAD), mean difference (meanAD), and peak value difference (PD) of signals in adjacent intervals can also be used as the characteristics of leak detection. If all the characteristics can be used together to detect the leak, it can greatly reduce the false alarm rate.

Hence, the maximum value of characteristics in normal conditions without a leak is used as the threshold, and the signals received by the acoustic sensors downstream are chosen for leak detection.

If the received value is larger than the threshold, a leakage will be detected; otherwise, the pipeline is in normal condition.

6.2.4 Detection of Leak Acoustic Signal in Buried Gas Pipe Based on the Time–Frequency Analysis

Min-Soo Kim, Sang-Kwon Lee*

In this paper, the cut-off frequencies of acoustic modes and dispersive phenomenon for acoustic wave propagation are researched based on wave equation analysis, experimental work, and a simulation by a commercial package (SYSNOISE).

The object of this paper is to identify the characteristics of this dispersive acoustic wave through analysis of the cutoff frequency by using the time-frequency method experimentally and BEM (boundary element method) theoretically for the development of an experimental tool to analyze the leak signals in the steel pipe.

The time-frequency method is a better analysis method than the PSD in order to identify the cutoff frequencies of acoustic modes and to analyze the dispersive wave.

The propagation from an acoustic wave in a duct used for a test is simulated by using the BEM. The SYSNOISE is used to analyze the acoustic wave in a duct [7].

6.2.5 Leakage Detection and Prediction of Location in a Smart Water Grid Using SVM Classification

Shailesh Porwal, Dr. S. A. Akbar, Dr. S. C. Jain

In this paper, the support vector machine technique is applied to detect leakage and predict the location in the water distribution and pipeline network of CSIR-CEERI, Pilani. In order to improve the accuracy and pinpoint the leakages, both the pressure values at junctions and flow values at pipelines are extracted from the EAPNET tool and used as a feature for analyzing through SVM to identify leakages rather than only the pressure value. EPANET is a simulation tool for water hydraulics and pipeline networks [8]. EPANET simulates the water hydraulic behavior of the water distribution and pipeline network that can be run over a desired time period and generates hydraulic parameters like pressure and flow at all points for that period of time. SVM techniques are supervised learning algorithms that use the learning datasets to train the model and further predict the future values using the trained model. SVM tries to obtain an optimal separation hyperplane that separates the different classes of learning data vectors to either sides of the hyperplane. SVM tries to maximize the margin distance between the support vectors.

6.2.6 Leakage Detection of a Spherical Water Storage Tank in a Chemical Industry Using Acoustic Emissions

Muhammad Sohaib, Manjurul Islam, Jaeyoung Kim, Duck-Chan Jeon, and Jong-Myon Kim.

In this paper, a quantitative assessment of acoustic emissions (AE) signals acquired from a spherical storage tank was performed. The proposed work is based on acoustic emission measurements and a machine learning technique to develop an intelligent fault diagnosis system [9]. This work can be used to find cracks in industrial components such as spherical tanks and pressure vessels. The main objective of this work is to perform data-driven structural integrity assessment of a spherical storage tank. The detection was performed using time-domain statistical features and a machine learning algorithm. The proposed method consists of (1) extraction of statistical features from the AE acquired from the spherical tank and (2) classification of the nonlinear data using an SVM. AE and electrical resistance measurements can detect early-stage as well as already existing damages in an object. AE testing has two main advantages: (1) No external energy is supplied in the AE recording; instead, the sound generated within the object is recorded. (2) AE testing can record dynamic changes in material, giving it the critical ability to discern growing and stagnant defects. These advantages under real-world conditions make acoustic emission testing a favorable choice to acquire data regarding tank structural integrity. The methodology can be divided into two main steps: (1) Extracting statistical features from the AE signals acquired from the spherical tank under normal and cracked states. (2) Feeding the extracted features to SVM to identify whether each input instance belonged to the normal spherical tank or the cracked state.

6.3 Methodology/Proposed Work

6.3.1 Machine Learning in Leak Detection

Data science and sophisticated data analysis are technologies that enable development of low-cost customization solutions. On that basis, a leak detection system in a pipeline employing an engine-learning methodology is discussed in this chapter. The methodology used to identify leakage was based on artificial intelligence, a framework for machine learning to optimize the pipe energy, along with an approach for anomaly identification [10]. Although the machine learning model used is a simple parametric linear regression and this technique is well known in the artificial intelligence domain, its competitive differential is the use of an open-source machine learning platform to implement them, which allows clients to have a customized model instead of using costly instrumentation with embedded systems. Various machine learning models were already used in previous works in this field. Machine learning models are used to check trends in the system in real time; in our system, we used the open-source machine learning library Scikit-learn to make it easy for future upgrades.

6.3.2 Importance of Sensors and Sensors Used

There are numerous elements that can effect leaks in plants and terminals, and it is difficult to determine what causes the abnormal behavior that led to the leak. Sensors are used to monitor the pressure density, temperature, and flow of pipelines. A sensor is a device that measures physical quantities and converts them to digital or analog signals. Temperature, length, force, and pressure are all examples of these quantities. The signal is typically electrical, but it can also be optical, as shown in Figure 6.1. There are several approaches for detecting leaks in a system. While different sensors are utilized for various leak detection systems, the sensors that our proposed system uses are mentioned next.

6.3.2.1 Pressure Sensor

A pressure sensor is a device that consists of a pressure-sensitive element that determines the actual pressure applied to the sensor and other components that turn this information into an output signal. The common unit of pressure is Bar(bar), Pascal(pa), N/mm^2, or Psi.

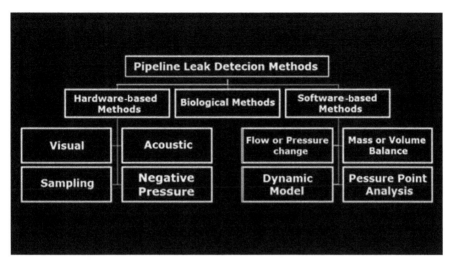

FIGURE 6.1
Pipeline leaks detection methods.

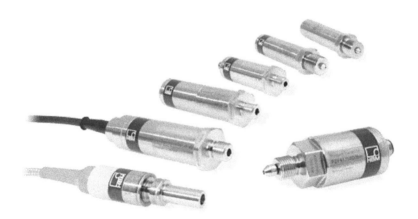

FIGURE 6.2
Image of pressure sensor. (HBK Company.)

Some examples of such sensors are potentiometric, inductive, capacitive, piezoelectric, strain gauge, variable reluctance, and barometric pressure sensors (Figure 6.2).

A pressure sensor is used in a gas leak detection system to provide an accurate reading of the pressure exerted on the pipeline during gas flow. The operator can manage the pressure of the gas using a pressure controller or valves. Pressure is one of the attributes whose flow density can aid in the detection and localization of leaks. The pressure sensor readings can be synced and stored in the form of data points, the values of which may be utilized for further analysis of the flow of gas in the pipeline during leakage and no leakage. In our model we used barometric pressure sensors of APG/AG3 Series with a range of 0–25 to 0–1,000 kpa. Our proposed system employs a threshold value to limit the value of pressure flow, such that if the pressure inside the pipeline exceeds the threshold value, an alarm will sound, signaling that a leak has been found.

FIGURE 6.3
Temperature sensor.

6.3.2.2 Temperature Sensor

A temperature sensor is a type of electronic device that senses the temperature of its surroundings and turns the measured data into electronic data in order to record, monitor, and communicate temperature changes. Temperature sensors have various types of applications, such as in computers, cooking appliances, automobiles, medical devices, etc.

Example: thermocouple, the RTD, the thermistor, semiconductor sensors, etc. (Figure 6.3).

Temperature sensors are used in leak detection systems to detect early the difference between the temperature inside and outside of the gas pipeline. In our system we are using the LM35 temperature sensor.

Features of a temperature sensor:

1. Suitable for remote applications
2. Low cost due to wafer level trimming
3. Operates from 4 to 30 volts
4. Lower heating issues
5. LM35 connects easily
6. Higher accuracy in values

6.3.2.3 Proximity Sensors

A proximity sensor is used to detect the presence of an object (commonly referred to as *target*). Proximity sensors are mainly used in smartphones, self-driving cars, industrial plants and terminals, anti-aircraft missiles, etc. The two most commonly used proximity sensors are the inductive proximity sensor and the capacitive proximity sensor. An inductive proximity sensor is used to detect metal targets, because it uses electromagnetic fields. The inductive characteristics of metal changes during contact with an electromagnetic field, and the object can be sensed at a larger or lesser distance based on how responsive the metal is (Figures 6.4 and 6.5).

FIGURE 6.4
Inductive proximity sensor.

FIGURE 6.5
Capacitive proximity sensor.

In contrast, capacitive proximity sensors are not confined to metallic objects. These proximity sensors are able to detect anything that can conduct an electrical charge. In liquid-level sensing, capacitive sensors are often utilized. Capacitive sensors' possible targets include, but are not limited to, glass, plastic, water, wood, metals, and a variety of other materials. In our system, a proximity sensor is utilized to detect human activity in the room where the pipeline is placed or where the possibility of a leak is greatest.

6.3.3 How Does It Work?

Gas leakage is one of the burgeoning topics right now across the globe. It has affected the economy and human resource to a great extent. A significant incident was the Vizag gas leakage where 11 people lost their lives and more than 1,000 people got sick after being exposed to the gas. Minor gas leaks happen every day across the globe that represent both direct and indirect risk within the health, environment, and safety aspects. Even smaller gas leaks are undesirable as they represent the risk of escalation that often leads to shut down or reduced production.

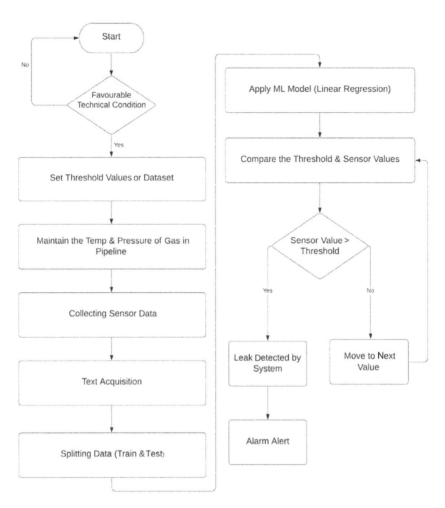

FIGURE 6.6
Flow chart of proposed model.

As a result, we developed a dependable system that uses machine learning and sensor data to identify leaks without the need for human intervention. Our suggested system has been tested on a hydrocarbon gas such as methane under temperature and pressure conditions. The dataset used contains the parameter of varying temperature and pressure of the underground pipeline in the city of Boston (dataset of 3 years). The flow of the system is shown in Figure 6.6.

The linear regression in the Scikit-learn package was used to evaluate the study hypothesis and the machine learning model. The system's key components include sensor data (methane gas), a machine learning model, and false alarm reduction. We carried out our research based on the paper titled "Machine learning and acoustic method applied to leak detection and location in low-pressure gas pipelines" proposed by Rodolfo Pinheiro da Cruz, which uses acoustic method to detect the leaks in the gas pipelines. Our finding contradicts earlier studies in that it is not dependable for long-distance pipelines, and the effectiveness of acoustic sensors decreases as pipeline distance increases. We proposed a work based on machine learning, in which we used a linear regression using the Scikit-learn library to figure out the leakage and identify the location of the leakage in the system to be able to warn construction site staff or the company about the leakage and take immediate action to avoid gas leakage and eliminate its causes. Primary gas detection will generate an alarm when sensing a given amount of gas in order to notify that the area might be explosive. However, it is necessary that small leaks are detected as well. Regression analysis is a form of predictive modeling that is useful to predict the output (i.e., dependent variable) of the system from the given input (i.e., independent variable) of the system. We chose regression as our primary method because it deals with continuous data and the significant dependent variable is numerical, as we use the sensor data where the value generated by the sensors are numeric. This helps us evaluate the output in real time. Since linear regression operates on continuous variables, it can evaluate the value in real time, and it is more effective than logical regression in gas detection (Figure 6.7).

The dataset above contains the temperature and pressure of the gas (methane, CH_4) at a specific moment, as shown by the columns FIRST_D and LAST_D. We must decide the regression line for this value in order to categorize our results. We used the training split method to split the dataset into subsets, it makes a random partition of the dataset, mostly 70% Training and 30% Test. We check the Test and Training values using a scatter map as we convert the values of the dataset into its associated Training and Test arrays (Figure 6.8).

```
      Entry_no    FIRST_D     LAST_D   Temperature      Pressure  Leak_Category
0            0   20150410   20160129     32.705980    -96.826468            Low
1            1   20150408   20160205     32.707543    -96.815914         Medium
2            2   20160119   20160129     32.709131    -96.843914            Low
3            3   20150408   20160204     32.713180    -96.817024            Low
4            4   20160104   20160127     32.713179    -96.817739            Low
[32.7059803 32.7075427 32.7091308 32.7131804 32.713179  32.7167425
 32.7177716 32.7196163 32.7199089 32.7200955 32.7199865 32.7204163
 32.7203876 32.7203981 32.7205424 32.7217155 32.7225301 32.7235265
```

FIGURE 6.7
A sample dataset of temperature and pressure of the gas.

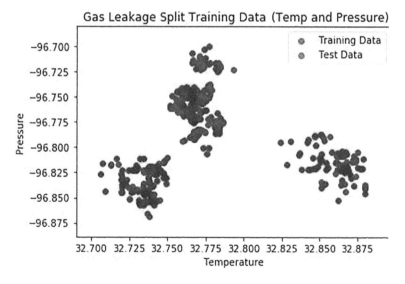

FIGURE 6.8
Plot showing the regression line of alert alarm and the dots representing the data points of the dataset.

(Continued)

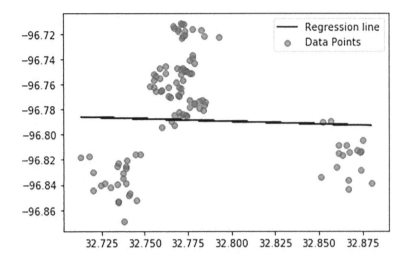

FIGURE 6.8 (*CONTINUED*)
Plot showing the regression line of alert alarm and the dots representing the data points of the dataset.

```
-96.8085408 -96.8152496 -96.8213669 -96.8522828 -96.8086232 -96.8065666
-96.82237   -96.8165819 -96.808675  -96.8366762 -96.8036441 -96.8327526
-96.8224265 -96.8086703 -96.8198887 -96.8170415 -96.808647  -96.8223613
-96.836089  -96.8250057 -96.8439054 -96.8049572 -96.8066631 -96.8140624
-96.8181415 -96.8198528 -96.8227535 -96.8262231 -96.830645  -96.8139964
-96.8123091 -96.8382349 -96.808529  -96.8139381 -96.8285118 -96.8129588
-96.811141  -96.8042053 -96.8195067 -96.8211133 -96.8211009 -96.811163
-96.8369154 -96.8386889 -96.8397432 -96.8409689 -96.8453749]
Regression Score 0.9814411369789104
```

FIGURE 6.9
Output regression score.

TABLE 6.1

Accuracy

Classifier	Accuracy (%)
SVM	86
KNN	88
Linear regression	98

The regression line depicted in the graph is the line along which the leakage alert will be displayed; we present our data in a scatter plot, which is nothing more than a display of our dataset's data points. We check the regression score of the system using the available function provided by the linear regression library (reg. score). The higher the score, the more effective the model is at detecting leaks, the greater the health or precision of the model, and the lower the fault alarm (Figure 6.9 and Table 6.1).

Our system has a regression score of approximately 98%; the higher the regression score, the more efficient our system is at detecting leakage. A regression scores close to 1.0 suggests positive outcomes and greater precision.

6.4 Linear Regression Algorithm

Linear regression is a supervised machine learning algorithm. It is one of the most easy and popular statistical methods used for predictive analysis. It is used to predict dependent variables (output values) based on independent variables (input values). The main objective of linear regression is to draw the best line so that the error between predicted value and real values is minimum [11].

6.5 Scikit-Learn Library

Scikit-learn Library, or Sklearn Lib, is one of the popular machine learning libraries. Scikit-learn is a python library that provides supervised and unsupervised machine learning algorithms for machine learning developers.

It provides tools for

- 1. > Regression
- 2. > Classification
- 3. > Preprocessing
- 4. > Clustering, etc.

Scikit-learn library is used in the ways shown in Figure 6.10.

Along with Scikit-learn, Pandas, NumPy, and Matplotlib are used.

6.6 Conclusion

This chapter presents the concept of finding a leakage in a pipeline system using machine learning algorithms. The linear regression method used on the data parameters like temperature and pressure of the gas shown in Figure 6.6 gives us the highest accuracy among all the other models, which is approx. 98.0%; the location can also be detected using this approach (linear regression method). In conclusion, various works in this field have also been compared to show the efficiency of our proposed system. The system is completely automated so that no human is harmed or injured during leak detection. The data processing carried out by the model and method is widely used to replace the human operator in the role of monitoring signals to alert the scale of the leakage and to notify the workers about the incident, and this technique could be extended to track natural gas pipelines, as well as industrial and residential gas pipelines, to ensure safe operation and prevent

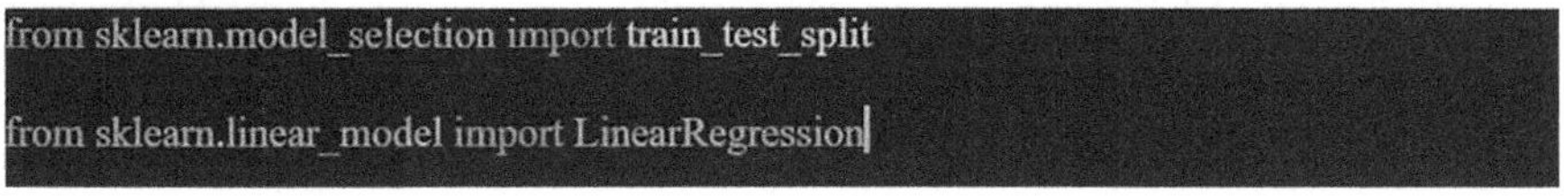

```
from sklearn.model_selection import train_test_split

from sklearn.linear_model import LinearRegression
```

FIGURE 6.10
Import linear regression.

serious human injury. Future work on this system should focus on real-time data with the goal of saving human life. Creating a mobile application so that gas leakage can be monitored from anywhere on the go, with the ability to operate the temperature and the flow of pressure of gas by a single click on a mobile screen. Sensors can be used to automatically close the valves of the portion where the pipe is damaged. We can use deep learning to train big datasets for higher performance and better results. Convolutional neural network (CNN) models can be trained to detect leakage and mark location using IR images. CNN has the advantage in dealing with 2D images as it uses convolution of image and generalizes it. Gas leakage detection will become easy when technologies like machine learning and artificial intelligence are used.

References

1. R. P. da Cruz, F. V. da Silva, and A. M. F. Fileti, "Machine learning and acoustic method applied to leak detection and location in low-pressure gas pipelines," *Clean Techn. Environ. Policy*, vol. 22, no. 3, pp. 627–638, Apr. 2020, DOI: 10.1007/s10098-019-01805-x.
2. M. Zhou, Z. Pan, Y. Liu, Q. Zhang, Y. Cai and H. Pan, "Leak detection and location-based on ISLMD and CNN in a pipeline," *IEEE Access*, vol. 7, pp. 30457–30464, 2019, DOI: 10.1109/ACCESS.2019.2902711.
3. J. Cho, H. Kim, A. L. Gebreselassie, and D. Shin, "Deep neural network and random forest classifier for source tracking of chemical leaks using fence monitoring data," *J. Loss Prev. Process Ind.*, vol. 56, pp. 548–558, Nov. 2018, DOI: 10.1016/j.jlp.2018.01.011.
4. M. Zadkarami, A. A. Safavi, M. Taheri, and F. F. Salimi, "Data-driven leakage diagnosis for oil pipelines: An integrated approach of factor analysis and deep neural network classifier," *Trans. Inst. Meas. Control*, vol. 42, no. 14, pp. 2708–2718, Oct. 2020, DOI: 10.1177/0142331220928145.
5. R. B. Santos, E. O. de Sousa, F. V. da Silva, S. L. da Cruz, and A. M. F. Fileti, "Detection and on-line prediction of leak magnitude in a gas pipeline using an acoustic method and neural network data processing," *Braz. J. Chem. Eng.*, vol. 31, no. 1, pp. 145–153, Mar. 2014, DOI: 10.1590/S0104–66322014000100014.
6. L. Meng, L. Yuxing, W. Wuchang, and F. Juntao, "Experimental study on leak detection and location for the gas pipeline based on acoustic method," *J. Loss Prev. Process Ind.*, vol. 25, no. 1, pp. 90–102, Jan. 2012, DOI: 10.1016/j.jlp.2011.07.001.
7. M.-S. Kim and S.-K. Lee, "Detection of leak acoustic signal is buried gas pipe based on the time-frequency analysis," *J. Loss Prev. Process Ind.*, vol. 22, no. 6, pp. 990–994, Nov. 2009, DOI: 10.1016/j.jlp.2008.08.009.
8. S. Porwal, S. A. Akbar, and S. C. Jain, "Leakage detection and prediction of location in a smart water grid using SVM classification," in *2017 International Conference on Energy, Communication, Data Analytics and Soft Computing (ICECDS)*, Chennai, Aug. 2017, pp. 3288–3292, DOI: 10.1109/ICECDS.2017.8390067.
9. M. Sohaib, M. M. M. Islam, J. Kim, C. Jeon, and J. Kim, "Leakage detection of a spherical water storage tank in a chemical industry using acoustic emissions," *Appl. Sci.*, vol. 9, Jan. 2019, DOI: 10.3390/app9010196.
10. E. Oliveira, M. Fonseca, D. Kappes, A. Medeiros, and I. Stefanini, "Leak detection system using machine learning techniques," *6th International Congress on Automation in Mining*, At Santiago, Chile, 2018.
11. S. Suryawanshi, A. Goswami and P. Patil, "Email spam detection: An empirical comparative study of different ML and ensemble classifiers," *2019 IEEE 9th International Conference on Advanced Computing (IACC)*, Tiruchirappalli, India, 2019, pp. 69–74, DOI: 10.1109/IACC48062.2019.8971582.

7

Two New Nonparametric Models for Biological Networks

Deniz Seçilmiş
Stockholm University

Melih Ağraz
Brown University

Vilda Purutçuoğlu
Middle East Technical University

CONTENTS

7.1 Introduction

The developing technology yields very promising improvements in scientific research area, rendering the raw biological data suitable for several types of statistical and computational techniques. One of the emerging areas is to infer biological networks as accurate as possible in order to reveal the interactions among any kind of subcellular parts of organisms, i.e., genes, proteins, etc. Representing the biological systems as networks can be regarded as one of the important issues since the biological incidents mostly are not easily visible [1]. Biological systems are very organized and everything is in order. However, in some conditions, specific differences may occur in these organized systems. In such situations, identification of the regular system as well as exhibiting the system under the disease conditions help researchers to determine the cause of the situation, solve the problems, and find alternative pathways to render the broken system organized again. There are many

DOI: 10.1201/9781003164265-7

other reasons to represent the biological systems as networks, such as for drug design. If the interactions among the molecules can be identified, researchers can identify which molecule regulates (activates or suppresses) the other(s), and then based on the pathway, the selection process of the drug target becomes much clear. The examples can be extended, but the general approach remains the same. Currently, it is possible to find various techniques for the modeling of such networks. The Boolean [2], ordinary differential equation (ODE) [2,3], and stochastic [2,4,5] approaches are the three main network modeling types. Among them, Boolean is based on the on/off position of the system and it is basically used for the description of new systems. The ODE approach is the model which represents the steady-state behavior of the system and it needs more detailed knowledge about the network with respect to the Boolean model [2]. Finally, the stochastic model is the most complex model for the description of biological networks since it is able to present the random nature of the system under continuing time [1,2]; whereas, as it requires detailed measurements about each system element, i.e., genes or proteins, currently it can be applicable for a limited number of biological networks. Hence, among these models, the ODE type is the most common choice since majority of the data is more suitable for this type of description [1,2].

However, since parameter estimation in ODE can be problematic for high-dimensional networks as it applies numeric solutions, its probabilistic alternative, called the Gaussian graphical model (GGM), has become popular [6]. This method shortly defines the states of the system via the multivariate normal density when the states are described by the lasso regression. In the inference of GGM, the model finds the entries of the precision matrix Θ, i.e., the inverse of the variance–covariance matrix, as it has a direct link with the regression coefficient in the lasso regression, and various alternatives are suggested to estimate Θ. Meinhaussen and Bühlmann [7] propose the neighborhood selection method; Yuan et al. [8] use the penalized likelihood method; Banerjee et al. [9] suggest both the block coordinate descent algorithm and the Nesterous first-order method. Then, Drton et al. [10] apply the SINful approach to control the incorrect edges in the estimated signal, and Friedman et al. [11] apply the coordinate descent algorithm by using the graphical lasso. The lasso approach is also inferred by the penalized maximum likelihood function under both L_1-norm and L_2-norm and its alternatives, which are named fussed lasso [12], adaptive lasso [13], and elastic lasso [14]. Finally, Ravikumar et al. [15] implement the logistic regression under an L_1 constraint to infer the parameter, i.e., the edges between a system's elements. In all these methods, as the inference of the model is still challenging for high percentage of sparsity in Θ, the model is further extended by the Gaussian copula [16] and is called as the Gaussian copula graphical model [17,18]. In the estimation of this model, the Bayesian approach can be applicable as the Bayesian algorithm is more suitable for high-dimensional inferences. Here, the reversible jump Markov chain Monte Carlo (RJMCMC) method [17,19] and the birth-and-death algorithm [20] are implemented. Furthermore, a semiparametric approach, named as the non-normal SKEPTIC algorithm [21] is suggested to decrease the computational demand of fully parametric models and another alternative is proposed by Ayyıldız et al. [22], called the lasso-based MARS, under the nonparametric framework. This model is free from any distributional assumption, resulting in more flexibility in the description of the states, and it can deal with the highly correlated sparse data. Due to the advantage of this nonparametric model over parametric GGM, in this study, we suggest both the random forest algorithm (RFA) and an extended version of lasso-based MARS since these algorithms were originally developed for highly correlated datasets without any strict distributional assumption.

Indeed, the RFA has common applications in clustering large and correlated datasets [23]; whereas it has not been used yet as an alternative nonparametric modeling in the

construction of biological networks. In this study, by detecting every pair of genes to identify motifs, modules and networks, we choose RFA since it is one of the main alternatives
of generalized additive models [23] under the distribution-free approaches. Within this
type of model, we can also consider the classification and regression tree (CART) method
[24,25]. However, since RFA is the extended version of CART, it is preferred for the deterministic construction of biological networks. In this study, we perform RFA on different
types of normal and non-normal datasets and measure the accuracies of estimates under
distinct measures such as precision, F-measure, and false positive rate by using Monte
Carlo studies. We also evaluate the performance of RFA under real biological datasets,
whose true links can be validated from the known literature about the systems. On the
other hand, the MARS approach is a well-known nonparametric method to model data in
various fields from time series [26] and voice data [27] to the diagnosis of breast cancer [28]
and classification problems. In order to estimate the model parameters, MARS follows a
two-step strategy, namely, the forward selection and the backward elimination. The forward step builds an overfitted model and the backward elimination reduces the model
complexity to get the optimal model. Thereby, in this study, we also propose to extend
the MARS method, which is suggested as an alternative of GGM, by a lasso regression
with main effects of genes [22,29,30]. Here, as the novelty, we add this lasso-based MARS
model to the second-order interaction effect and evaluate the accuracies of results in two
real datasets. From comparative analyses via GGM, lasso-based MARS, and lasso-based
MARS with interaction terms, we observe that the suggested approach validates the biological knowledge with new findings about the systems. Hence, we consider that both RFA
and lasso-based MARS with interaction terms can be promising to describe the steady-
state activation of the biological networks under distinct dimensions and the distribution
of genomic observations.

Accordingly, in the rest of the chapter, we present the underlying outcomes as follows:
In Section 7.2, RFA, GGM, MARS, and lasso-based MARS with interaction models are
explained briefly. Then, the application of all proposal methods is presented in Section 7.3.
Finally, Section 7.4 summarizes with conclusions and presents our future directions.

7.2 Methods

7.2.1 Random Forest Algorithm

The RFA [23] is a procedure that applies classification or regression depending on the data
constituting factors or continuous variables, respectively. The classification method under
RFA is categorized under supervised machine learning techniques [31]. In the modeling of
protein–protein interactions, the classification plays an important role when the construction of the trees is done recursively in a binary form. In this classification, the idea is to
assign each observation to the correct subgroups whose elements are previously known.

In the standard classification trees, the branching process continues by splitting each node
with respect to the best groups among all variables. In the selection of the best groups, the
branching should go only with the measurements (mainly, generalization error, strength,
and correlation). However, it also goes with the noises of the measurements, causing the
overfitting problem during the calculation. One approach to overcome this challenge is
to select each tree independently from the previously chosen ones in the forests. Bagging

[23] is an algorithm that provides such a solution. Here, it constructs each tree indepen-dently by selecting a bootstrap sample of the datasets, and then it chooses the most voted tree for the prediction. Even though the bagging reduces the variance very effectively, it is not totally enough to discard the overfitting problem since it also causes bias during variance reduction. Therefore, adaptive bagging [23] is suggested as the improved version of bagging to reduce both the variance and the bias effectively. Additionally, it increases the accuracy by improving the estimates of the main measurements (generalization error, strength, and correlation) of combined ensembles of trees. Hence, it prevents the effect of the bagging error on the estimates of the measurements, which has a remarkable impor-tance in the inference of biological systems. However, the adaptive bagging algorithm is designed to work well on large datasets, which is not always possible when the data come from biological experiments or personal health records. In RFA, an upper boundary is gen-erated by limiting the generalization error (pe^*) in order to prevent the overfitting problem without requiring large datasets.

$$- P_{x,y}\left(P_\theta\big(h(x,\theta)=y\big) - \max_{(j=y)} P_\theta\big(h(x,\theta)=j\big) \right) < 0. \tag{7.1}$$

In Equation (7.1), $\max_{(j=y)} P_\theta$ ($h(x,\theta)=j$) denotes the maximum of all the probability val-ues among all the values of the classifier except for its value on the point y while $h(x,\theta)$ denotes the classifier for the random vector x. Since the main rule in RFA is to maximize the strength between nodes (in biological systems, proteins) with the lowest correlation, adaptive bagging is a useful basis due to its good estimation of the main measurements of strength, correlation, and the generalization error with an additional process for limit-ing the generalization error. In the calculation, it can successfully create an upper bound-ary to increase the accuracy of trees, modules (pathways), and the resulting networks. Furthermore, it splits each node by taking the best division among a subset of predictors randomly chosen at a specific node, rather than splitting among all variables, to prevent overfitting.

Moreover, by creating small communities, the construction of each tree is controlled by generating a representative random vector denoted by θ. Here, θ_k refers to the growth of the kth tree in the ensembles representing each of those small communities by the num-bers from 1 to k. In this way, k also represents the convergence of trees in the forest. The accuracy of a random forest is defined via the generalization error, the strength, and the dependence measure correlation. Hence, the generalization error (pe^*) of RFA is controlled by the following expression:

$$pe \leq \bar{\rho}\left(1-s^2\right)\big/s^2. \tag{7.2}$$

In Equation 7.2, the mean value of the correlation between random vectors θ and θ' is shown via ρ while θ' is the proposal tree in the next iteration. Here, s indicates the scalar vector of the strength of the small communities via $s=E_{x,y}mr(x,y)$. E implies the expectation between the random vectors x and y, while the margin function is denoted by $mr(.)$. Hereby, in this study, we suggest RFA as an alternative model for the deterministic description of the steady-state activation of biological networks. Since RFA is a nonparametric approach that does not deal with the variance–covariance matrices and the threshold values, the basis of RFA depends on whether there is an interaction between nodes or not. Therefore, a number of iterations is performed depending on the size of the data, and the nodes, i.e., proteins, are bound to each other iteratively by maximizing the strength while minimizing the correlation. Thus, in our calculation, initially, we bind the two closest proteins to each other with respect to their instance values from the confusion matrix of RFA as an output at the end of each RFA iteration, and then we continue this process either by completely binding two new proteins to each other or by binding an individual protein to the previously bound nodes. We repeat the underlying iterations until there are no related proteins left to bind. Accordingly, in the resulting matrix, all the proteins can be bound to each other. Furthermore, there can be individual proteins that are not interacting with any of the other proteins in the system. The latter system is called as a sparse network, which is highly common when the deal is the biological data.

7.2.2 Gaussian Graphical Model

The graphical model [32], as shown in Figure 7.1, is a popular tool for the sparse structure and supplies a graphical representation of random variables under their conditional independencies.

In this model, we suppose that $Y = (Y_1, Y_2,..., Y_p)$ is a p-dimensional multivariate normally distributed random variable with a p-dimensional mean vector μ, i.e., $\mu = (\mu_1, \mu_2,... , \mu_p)$ and a $(p\,p)$-dimensional covariance matrix Σ, whose entries are $\sigma_{i\,j}$'s ($i, j = 1,.... , p$) as shown below:

$$\mu = \begin{pmatrix} \mu_1 \\ \vdots \\ \mu_p \end{pmatrix}, \quad \Sigma = \begin{pmatrix} \sigma_{1,1} & \cdots & \sigma_{1,p} \\ \vdots & \ddots & \vdots \\ \sigma_{p,1} & \cdots & \sigma_{p,p} \end{pmatrix} \tag{7.3}$$

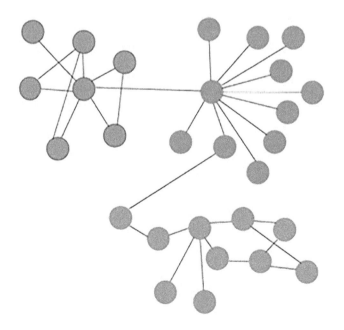

FIGURE 7.1
Simple representation of a network with 28 nodes via an undirected graph.

So the model constructs a least absolute shrinkage regression, also named as the lasso regression [33], to explain the interaction between the system's elements, which are genes in this study.

$$Y_p = \beta Y_{-p} + \varepsilon. \tag{7.4}$$

In Equation 7.4, Y_p denotes the state of the pth node, Y_{-p} shows the states of all nodes except the pth node, β refers to the regression coefficients, and ε is the error term, which is independent and normally distributed.

In GGM, we are interested in the estimation of the inverse of the covariance matrix, which is also called the precision matrix, $\Sigma^{-1} = \Theta$. Because this matrix reflects the conditional independent structure of the system's elements under an undirected graph, as seen in Figure 7.1. Accordingly, β in Equation 7.4 can be described in terms of the entries of θ, i.e., θ_{ij} ($i, j=1,\ldots, p$), as presented below:

$$\beta = -\Theta_{-p,p}/\Theta_{p,p}. \tag{7.5}$$

In Equation 7.5, the estimated interaction between two nodes in a system is explained by the associated entries of the precision matrix. In this expression, Y_p and Y_j ($j=1, \ldots, p$) are conditionally independent when $\beta=0$ [34] and the optimal estimate of β is found via

$$\hat{\beta}(\lambda) = \arg\min_{\beta} \left\{ \frac{\left\| Y - Y_p \beta \right\|_2^2}{n} + \lambda \|\beta_1\| \right\} \tag{7.6}$$

In Equation 7.6, $\|\cdot\|_1^1$ and $\|\cdot\|_2^2$ stand for the L_1-norm and the L_2-norm of the given values, respectively. As seen in the objective function of β, the estimated parameters can be found using different optimization methods under the high-dimensional and sparse θ. In this study, we perform the graphical lasso (glasso) [35] due to its simplicity, whereas, the inference of GGM in our analyses can be also implemented by other estimation methods.

7.2.3 Multivariate Adaptive Regression Splines

MARS [36] is a regression model used to identify linear and nonlinear effects as well as interactions between covariates by means of piece-wise nonlinear models. In MARS, the nonlinear relation between the response variable y and the predictor x's is described as

$$y = \beta_0 + \sum_{m=1}^{M} \beta_m H_m(x) + \varepsilon \tag{7.7}$$

in which β_0 denotes the intercept term and β_m is the regression coefficient as used earlier. M presents the total number of parameters, ε is the random error term, and H_m shows the spline basis function (BF), as seen in Equation 7.9. BFs are defined by

$$(x-t)_+ = \begin{cases} (x-t) & \text{if } x > t \\ 0 & \text{otherwise} \end{cases} \quad \text{and} \quad (t-x)_+ = \begin{cases} (t-x) & \text{if } x > t \\ 0 & \text{otherwise.} \end{cases} \tag{7.8}$$

In the above expression, t represents the knot and physically implies the breaking point of the spline function, as simply drawn in Figure 7.2. Here, the spline basis functions H_m are defined via

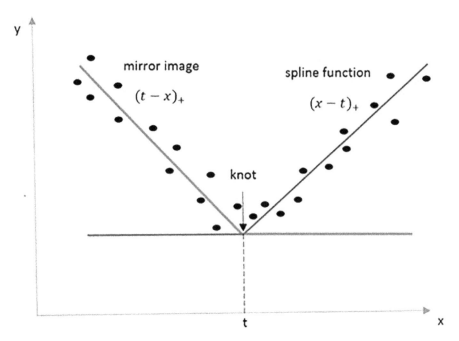

FIGURE 7.2
Simple representation of the MARS model.

$$H_m(x) = \prod_{k=T}^{K_m} \left[\max\left(S_{k,m} \left(x_{v(k,m)} \right) - t_{k,m}, 0 \right) \right], \tag{7.9}$$

where K_m shows the number of truncated functions in the mth BF, $x_{v(k,m)}$ stands for the input variable corresponding to the kth truncated linear function in the mth BF, $t_{k,m}$ corresponds to the knot value for H_m, and $s_{k,m}$ takes the value +1 for the basis function and −1 for the mirror image.

In the detection of the optimal model, MARS performs two steps sequentially. These are the forward selection step and the backward elimination step. In the forward selection step, a large model with many BFs is obtained and this procedure can cause overfit of the data. On the other hand, in the backward elimination step, some of the predictors in the forward selection part are discarded to reduce this overfit. In the elimination, the generalized cross validation (GCV) value [36,37], whose formula is given in Equation 7.10, is the major criterion.

$$GCV = \frac{\sum_{i=1}^{N} \left(y_i - \hat{y}_t \right)^2}{\left(1 - M/N \right)} \tag{7.10}$$

Here, y_i is the ith response value, $\hat{y}_i$ represents the corresponding fitted response value, and N shows the number of samples. Moreover, M denotes the effective number of maximum BFs and is found from the expression $M = r + cK$. Here, c is the cost in the optimization of the basis function and the smoothing parameter of the model [38]. This value is typically equated to $c = 3$, whereas, if the model is an additive model, it is set to $c = 2$ [35]. On the other hand, the selection of the penalty constant can also be found by different model selection criteria, rather than GCV.

For instance, in the study of Kartal-Koc and Bozdogan [39], the selection of the optimal MARS model is done via the information theoretic measure of complexity (ICOMP) [40,41] under the multivariate normally distributed data. But, a comprehensive performance of other model selection criteria [42], besides ICOMP, has not been evaluated yet for different types of biological networks.

In the lasso-based MARS, as the alternative of GGM, we construct a regression model for each node against all remaining nodes similar to the lasso regression, as shown in Equation 7.4. From previous analyses [22], it has been shown that since the model building of the MARS strategy is similar to the stepwise multiple linear regression due to the application of basis functions, it can have more flexibility than GGM to capture the nonlinear structure of the data, resulting in better fitting to describe the biological networks, which include high nonlinearity because of the scale-freeness. Moreover, the MARS model is particularly designed for highly correlated data without any distributional assumption. Therefore, when the MARS model is adapted to construct the biological networks, it has better performance than GGM, especially, for high-dimensional and non-normal datasets. Hereby, in this study, besides the current advantages of the lasso-based MARS over GGM, we also include the second-order interaction terms in our model due to the fact that its physical structure can be thought as the feed-forward loop in the biological networks [43]. Because such a nonlinear description cannot be represented under GGM as it is strictly based on the normality assumption which implies the linear relationships between the genes.

We can also show the benefit of the application of MARS with/without interaction effects in a small system. For instance, let's assume that we have a system with four nodes and each node in this system is estimated by the following sets of lasso equations without interactions and with interactions by MARS in Table 7.1.

Then, if we describe these sets of equations via the estimated edges in the system, we can report that the first node y_1 has connections with node 2, y_2 has edges with nodes 1 and 4, y_3 is autoregulated, and y_4 is bounded with nodes 1 and 2 when the interaction effect is not added into the model. Whereas, if it is included in the model, then, y_1 has connections with y_2, y_3, and y_4; y_2 is connected with y_3 and y_4; y_3 has edges with y_2 and y_4; and y_4 is bounded with y_1 and y_3. Therefore, the associated adjacency matrix of MARS without interactions (Θ_{woi}) and MARS with interactions (Θ_{wi}) can be denoted as below:

$$\Theta_{woi} = \begin{bmatrix} 1 & 1 & 0 & 1 \\ 1 & 1 & 0 & 1 \\ 0 & 0 & 1 & 0 \\ 1 & 1 & 0 & 1 \end{bmatrix} \quad \text{and} \quad \Theta_{wi} = \begin{bmatrix} 1 & 1 & 1 & 1 \\ 1 & 1 & 1 & 1 \\ 1 & 1 & 1 & 1 \\ 1 & 1 & 1 & 1 \end{bmatrix} \tag{7.11}$$

TABLE 7.1

Representation of a System with Four Nodes without and with Interactions by MARS

Description of System	Lasso Equations
Lasso-based MARS without interactions	$y_1 = 2y_2$ $y_2 = y_1 + y_4$ $y_3 = y_2 \times y_4$
Lasso-based MARS with interactions	$y_1 = y_2 + 2y_3$ $y_2 = 3y_3 + y_4$ $y_3 = y_2 \times y_4$ $y_4 = y_1 \times 2y_3$

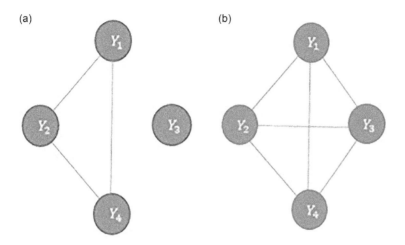

FIGURE 7.3
Estimated network for a system with four genes represented in Table 7.1 by MARS (a) without interaction and (b) with interaction effects.

where the columns and the rows show the name of the genes from 1 to 4, sequentially. In other words, we consider that the elements of interactions also imply the pairwise relations between each other and between the response, separately. We also represent the graphical view of these matrices in Figure 7.3.

7.3 Application

7.3.1 Application via Random Forest Algorithm

In this study, in an attempt to evaluate and compare the results of the random forest algorithms (RFA), firstly, the true positive (TP), true negative (TN), false positive (FP), and the false negative (FN) numbers are calculated, and based on these values, the precision, recall, F-measure, false positive rate (FPR), and Matthew's correlation coefficient (MCC) values are computed. The mathematical expressions of these measures are also denoted in Table 7.2.

7.3.1.1 Application via Simulated Data

In this study, to perform RFA under realistic structures of the biological networks, we produce different dimensional networks whose adjacency matrices indicate the scale-freeness, which is one of the main topological features of the biological systems and it is related to the connectivity of the nodes [44]. Accordingly, the scale-free system implies that the network is highly sparse and there are a few nodes, called hubs, that have high connections with other nodes and many nodes that have just a few links with others. Moreover, when the challenge is the biological network data, it is not always possible to approach it with the normality assumption. So, as we suggest a nonparametric model for overcoming the restriction of the normality in the description of the system, we need to detect whether

TABLE 7.2

The Expressions of Accuracy Measures Based on TP, TN, FP, and FN

Accuracy Measure	Formula
Precision	$\dfrac{TP}{TP + FP}$
Recall	$\dfrac{TP}{TP + FN}$
F-Measure	$2 \times \dfrac{\text{Precision} \times \text{Recall}}{\text{Precision} + \text{Recall}}$
FPR	$\dfrac{FP}{TN + FP}$
MCC	$\dfrac{\sqrt{TP \times TN - FP \times FN}}{(TP + FP)(TP + FN)(TN + FP)(TN + FN)}$

our proposed nonparametric algorithm (RFA) provides equal outputs on the data far away from the normality assumption as it provides on the normally distributed data. Therefore, we apply the Gaussian copula with respect to exponential, student-t, and log-normal marginal distributions as well as the mixture margins, i.e., half normal-half exponential and half normal-half log-normal, in order to reveal the performance of the suggested algorithm under different joint distributions. In the selection of these marginal distributions, we consider that the exponential distribution can be used to express the measurement of the cell signal [45], the student-t distribution is one of the close alternatives of the normal density, and the log-normal distribution can be preferable if the logarithm of the raw data can be described by the normal density [46]. Then, we take fully unique margins and the mixture of two marginal distributions in order to comprehensively evaluate the performance of RFA under various scenarios. Afterward, we bind these margins via the correlation structure of the Gaussian copula, and compute our accuracy measures in Table 7.2 based on 1,000 Monte Carlo runs. Additionally, in our analyses, we run the model under 10, 20, 50, and 100 genes/proteins to detect the effect of the system's dimension on modeling, and also in all simulations, we set the number of observations per gene/protein to 20, which is particularly reasonable for high-dimensional systems.

On the other hand, in the application of RFA in R, first, we create a symmetric and an empty adjacency matrix whose row and column names are the protein labels. When RFA starts to proceed, it creates a confusion matrix consisting of instance values at the end of each run. Since our purpose is to have the maximum strength between two nodes with very few correlation, we accept the instance values in the confusion matrix as equivalent to the strength, and at each run, we aim to select the nodes having the maximum strength. After we choose the proteins from the confusion matrix, we turn back to the original data consisting of observations, which belong to the proteins, and then label them with a common name. Afterward, the new form of the data becomes exposed to another random forest run, and once again the couples with the maximum strength are chosen from the confusion and labeled together in the original observation data. This process is forced to continue until there is no remaining protein to bind. During this iterative procedure, at the end of each run, we record the protein names before labeling them together, recall the adjacency matrix that we have created at the beginning of the procedure, and fill the corresponding cell with 1. In constructing the network, we do not only cope with

TABLE 7.3

RFA Results Based on 1000 Monte Carlo Runs under Multivariate Normal, Student-t with Degree of Freedom 5, Exponential with Rate 5 and Log-Normal with Mean 10 and Standard Deviation 2 Distributed Data and 10, 20, 50, and 100 Number of Nodes in the System, Respectively

	Normal				Student-t	
Measures	10	20	50	100	10 20	50 100
Precision	0.58	0.57	0.59	0.63	0.01 0.02	0.00 0.00
Recall	0.37	0.36	0.34	0.34	0.01 0.02	0.00 0.00
F-Measure	0.46	0.44	0.43	0.44	0.01 0.02	NC NC
FPR	0.10	0.06	0.01	0.01	0.04 0.10	0.02 0.01
MCC	0.32	0.38	0.43	0.45	−0.03–0.08	−0.03–0.01
	Exponential				Log-normal	
Measures	10	20	50	100	10 20	50 100
Precision	0.02	0.01	0.00	0.00	0.00 0.00	0.00 0.00
Recall	0.02	0.01	0.00	0.00	0.00 0.00	0.00 0.00
F-Measure	0.02	0.01	0.00	0.00	NC NC	NC NC
FPR	0.18	0.09	0.02	0.02	0.15 0.09	0.03 0.01
MCC	−0.18	−0.09	−0.03	−0.01	−0.17–0.10	−0.03–0.02

NC implies noncomputable values.

the protein pairs, but also with the motifs and the modules. This occurs when a protein shares the maximum strength with a previously bound couple, goes and binds to them, and these new complex proteins constitute a structure together. In such cases, in order to find out which protein in the couple (or motif or module) the single one binds to, we check the first confusion matrix in which all the proteins stay single and search for the maximum strength among all possible combinations of those proteins. Finally, in the adjacency matrix, the cell with the labels of the chosen proteins is filled with 1. When the procedure ends due to no protein left to bind, the empty cells of the adjacency matrix are filled with zeros. In the end, the adjacency matrix becomes ready to be compared with the true precision matrix.

Accordingly, Table 7.3 indicates the outputs of the multivariate normal, student-t, exponential, and log-normal distributed data under 10, 20, 50, and 100 genes. From the results, it can be seen that the performance of RFA improves for high dimensions under normality, whereas under non-normal distributions, the accuracies decrease and they are irrelevant to the dimension of the systems. We observe similar conclusions from the findings of the mixture distributions as seen in Table 7.4.

7.3.1.2 *Application via Real Data*

In the assessment of RFA in real systems, we use two benchmark datasets. The first dataset is called cell signaling data consisting of 11 molecules to describe the phosphorylation of the proteins under various experimental conditions in human primary naive CD4+T cells that are measured on 11,672 red blood cells. Briefly, the dataset has a 11,672×11 dimension. These data are collected after a series of stimulatory cues and then, the inhibitory interventions with cell reactions are stopped at 15 minutes after the stimulation by a fixation in order to profile the effects of each condition on the intracellular signaling network [47]. Figure 7.4 shows a simple representation of this system. From the perspective of the

TABLE 7.4

RFA Results Based on 1000 Monte Carlo Runs under Non-normally Distributed Data Whose Margins Are Half Log-Normal with Mean 10 and Standard Deviation 2 and Half-Normal with Mean 10 and Standard Deviation 2 as well as Half Exponential with Rate 5 and Half Normal with Mean 10 and Standard Deviation 2, Respectively; and Whose Number of Nodes Are 10, 20, 50, and 100

	Normal/Log-Normal			Normal/Exponential			
Measures	10 20	50	100	10	20	50	100
Precision	0.20 0.04	0.00	0.00	0.10	011	0.00	0.02
Recall	0.22 0.05	0.00	0.00	0.11	0.11	0.00	0.03
F-Measure	0.21 0.04	NC	NC	0.11	0.11	NC	0.03
FPR	0.20 0.14	0.06	0.01	0.22	0.10	0.04	0.02
MCC	0.03-0.07	−0.05	−0.02	−0.10	0.01	−0.04	0.01

NC implies noncomputable values.

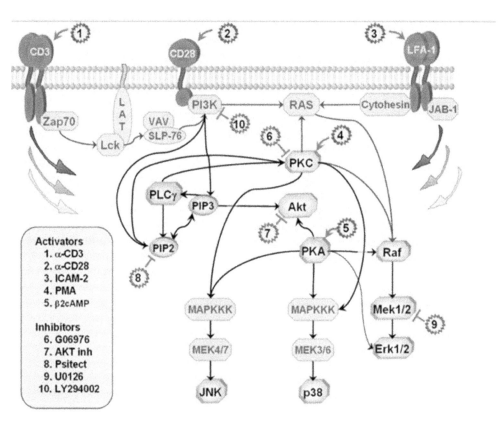

FIGURE 7.4
True graphical representation of the cell signaling network. (From the study of Sachs et al. [55].)

network topology, when the degree distribution is checked by measuring the correlation coefficient, the appropriate network topology definition for the cell signaling data can be said to be *random* [44] since the number of links per protein is almost the same for each protein, and the average of the total links and the number of genes is equal to this number.

TABLE 7.5

Accuracy Measures of RFA from the Cell Signaling Pathway

	RFA
Precision	1.00
Recall	0.04
F-Measure	0.09
FPR	0.00
MCC	0.17

NC refers to the noncomputable value

From the results in Table 7.5, it can be seen that RFA proves itself in the construction of the cell signaling pathway.

The second dataset is called as the human gene expression pathway data, which was gathered by Stranger et al. [48], and is described by Bhadra and Mallick [49] and Chen et al. [50]. This dataset is collected to measure the gene expression in the B-lymphocyte cells in people of Northern and Western European ancestry from Utah (CEU). The data are composed of 60 unrelated individuals for 100 probes. Briefly, the dataset has a 60×100 dimension. Here, the focus is on the 3,125 single nucleotide polymorphisms that are found in the 5 UTR (untranslated region) of mRNA (messenger RNA) with a minor allele frequency of 0.1. Since UTR of mRNA has an important role in the regulation of the gene expression, the inference of this system has been performed in the previous study [20] via the copula GGM. Different from the first real dataset, here the true precision of the gene expression pathway is unknown, leading us to not compute the accuracy measures for the comparison of RFA outputs and to not have the idea of the network topology. Therefore, for this human gene expression dataset, we run RFA by excluding the calculation of the accuracy measures based on TP, TN, FP, and FN, and instead, we only record the interactions that are found in the resulting precision matrices. The results exhibit that RFA can detect the new interactions as well as capture validated interactions based on the databases STRING and GeneMANIA, which exhibit the true structure of protein–protein interactions and gene interactions, respectively. Table 7.6 illustrates the interactions between molecules that

TABLE 7.6

The List of Interactions between Molecules That Are Recorded as Final Results of RFA from the Human Gene Expression Data, Where ** Indicates the Validated Interactions Based on Both STRING-DB and GeneMANIA-DB, and * Refers to the Validated Interactions Based on Only GeneMANIA-DB

Interactions	Molecule 1	Molecule 2
Biologically validated interactions		
	HMOX1	IL8[b]
	RPS4Y1 DDX3Y TNFRSF19	EIF1AY[b] KDM5D[b] LEPREL1[a]
New interactions		
Close localization based on STRING-DB	EPS8 G0S2 ABCC6	STEAP1 IL8 KDM5D
	MOXD1	LEPREL1
Others based on STRING-DB	EPS8	RGS13
	TCEAL2 STEAP1	F13A1 HLA-A

[a] Refers to the validated interactions based on only GeneMANIA-DB.
[b] Indicates the validated interactions based on both STRING-DB and GeneMANIA-DB.

are detected via RFA; and among these interactions, the ones between HMOX1 and IL8, RPS4Y1 and EIF1AY, and between DDX3Y and KDM5D are validated based on both the STRING and GeneMANIA databases (DB); while the interaction between TNFRSF19 and LEPREL1 is validated based on only the GeneMANIA database.

In order to exhibit the true situation of protein–protein interactions among selected proteins via RFA, we convert the probe IDs into protein IDs. Hereby, considering the localization of the proteins in Figure 7.5 prepared from the STRING-DB results, especially, EPS8 and STEAP1, and also G0S2 and IL8, ABCC6 and KDM5D, and MOXD1 and LEPREL1 seem close to each other, which can be considered as plausible new biological pathways.

Afterward, we display the true gene interactions among these captured nodes via RFA from the GeneMANIA database. Figure 7.6 illustrates the true gene interactions with co-localization and genetic interactions, in which dark edges refer to the co-localization of the genes, while gray edges among genes indicate the genetic interactions. Some of the nodes are observed close to each other, including EPS8 and STEAP1, ABCC6 and KDMD5, which are close in STRING-DB results, and STEAP1 and HLA-A, DDX3Y and KDM5D, which are close in GeneMANIADB findings. Hence, considering these gene interactions with co-expression based on GeneMANIA database, it can be said that RFA is strong enough to capture all these interactions from the data with no false positives in terms of co-expression.

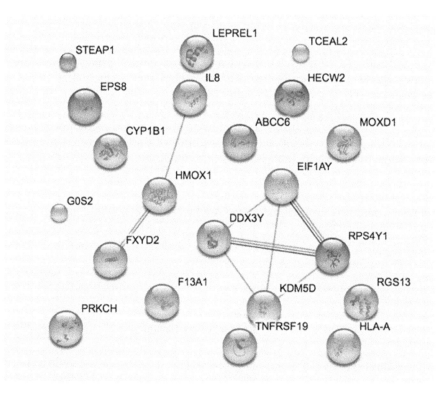

FIGURE 7.5
The true representation of the selected proteins and their interactions based on the STRING-DB at the end of RFA analysis. The smaller nodes correspond to the proteins whose 3D structures are unknown while the 3D structures of the bigger nodes are known.

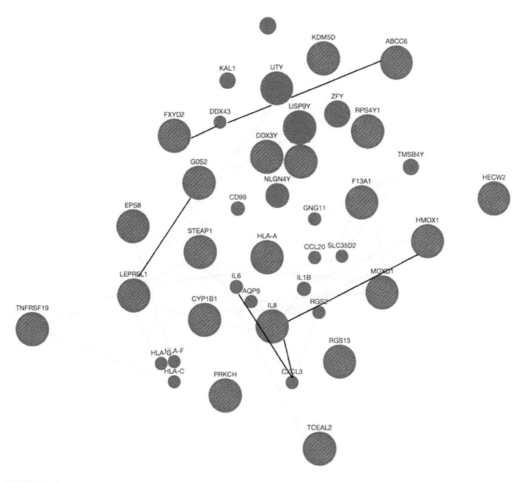

FIGURE 7.6
True gene interactions with co-localization and genetic interactions based on the GeneMANIA-DB. The dark edges refer to the co-localization of the genes, while other edges among genes indicate the genetic interactions.

7.3.2 Application via MARS and Gaussian Graphical Model

MARS is a promising alternative approach of GGM in the construction of the networks [22]. In this part, we perform this model with the interaction effects and make comparative analyses with GGM and MARS without interaction terms in order to assess the gain in this extended model. In the application of MARS with/without interaction effects, the inferences of precisions are constructed in such a way that every gene is sequentially assigned as a response and the remaining genes are used as covariates. For example, in the first iteration, the first column (Y_1) is chosen as response and the remaining is used as covariate (Y_{-1}), in the second iteration, the second column (Y_2) is assigned as response the variable and the remaining is used as covariate (Y_{-2}). These steps are followed until the last column is assigned as response. Therefore, MARS can only handle one response variable in every iteration. Then, under the estimated regression coefficients, we take significant covariates from each gene-specific model. Here, we accept that these entries imply the significant relations between the response gene and the associated predictor gene, resulting in an entry 1 in the adjacency matrix. Otherwise, we put 0 value in that entry. Furthermore, we apply the OR rules in modelling via MARS. This rule means that, if there is a connection in the entry of the adjacency matrix [*i, j*] and [*j, i*] entry (*i, j*=1, 2, … , *n*), we set 1 for both [*i, j*] and [*j, i*] entries in this matrix in both MARS with/without interaction models.

On the other hand, in the GGM analysis, the best penalty constant is chosen by consistent Akaike's information criterion (CAIC) and consistent Akaike's information criteria with Fisher information (CAICF) of Bozdogan [42] and information theoretic measure of complexity (ICOMP) criterion of Bozdogan [40,41]. Furthermore, the glasso package is used to estimate the precision matrix [51] and a threshold value of 0.1 is chosen arbitrarily to transform the precision matrix to a binary adjacency matrix. In other studies, a similar approach was adopted while choosing the threshold value and successful results were obtained [22]. While transforming the adjacency matrix, the values greater than the threshold value are assigned a 1 and the others are assigned a 0.

Hereby, in application, all GGM and MARS models with/without interaction effects are used in the same real datasets applied with RFA. As stated beforehand, the first data [47] are called the cell signaling data and show a small network having 11 phosphoproteins and phospholipids. In Figure 7.7, the estimated systems via all alternative models are shown. In this figure, it can be seen that the MARS model with interaction effects estimates biologically validated links and already infers the links found via MARS without interaction effects and GGM. For instance, the link between PKA and PKC proteins (protein numbers 8 and 9 in Figure 7.7), which is biologically validated in the study of Sim and Scott [52], is merely found via MARS with the interaction effect. Furthermore, the edge between PRAF and PMEK proteins (protein numbers 1 and 2 in Figure 7.7) is correctly inferred [53] in GGM and MARS with interaction effects. The link between PIP2-PKC proteins (protein numbers 4 and 9 in Figure 7.7) is not estimated via MARS without an interaction model, whereas it is inferred under the interaction model as biologically declared in the study of Kuo [54]. In addition, it can be seen that the relation between PRAF and PKC proteins (protein numbers 1 and 9 in Figure 7.7) is validated by MARS with interaction effects. This interaction can be verified by Figure 7.4 taken from Sachs et al. [47]. In addition, MARS with interaction can catch the relation between PLCG and PKC proteins (protein numbers 3 and 9 in Figure 7.7). This relation can also be validated by Figure 7.4.

In the second application with a real dataset, we use the large-scale human gene expression data as implemented via RFA. As previously described, the data consist of 60 unrelated individuals for 100 genes. In this part, we estimate the associated network via both alternatives of MARS and GGM as presented in Figures 7.8–7.10. From comparative analyses,

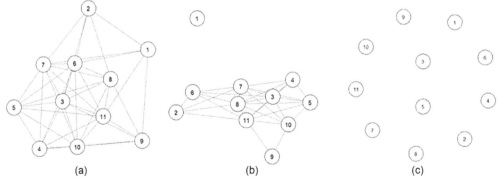

(a)   (b)   (c)

FIGURE 7.7 Estimated structure of the cell signal network via (a) MARS without interactions, (b) MARS with interactions and (c) GGM under ICOMP, CAIC, and CAICF criteria.

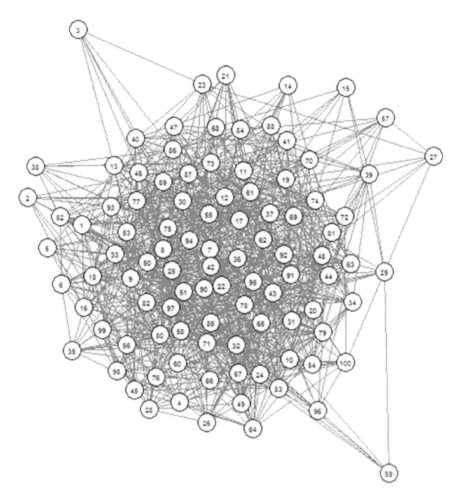

FIGURE 7.8
Estimated structure of the human gene interaction network via MARS without interaction effects.

it is found that both MARS models can capture biologically validated links and GGM cannot detect any interaction. Then, due to the failure of the estimation via GGM under this dimension, we also perform an approximate version of the glasso approach, which is particularly designed for such high-dimensional networks. Hereby, we infer the parameters of GGM as in the study of Zhou et al. [50] via the huge package in R. This method and its associated package basically estimate the structure of the graph from the multivariate Gaussian distribution by using a multi-step idea. This specific calculation is called Gelato (Graph estimation with Lasso Thresholding). Basically, Gelato infers the structure of the systems by two stages. In the first stage, an undirected graph is estimated via a threshold among the L_1-norm penalized regression functions, and in the second stage, the variance–covariance matrix and its inverse are estimated via their maximum likelihood estimators. The inference of the interactions, i.e., the edges of the graph, is performed as in the study of Meinshaussen and Bü"hlmann [7], which implements the lasso regression to obtain the vector of coefficients. In this computation, the large number of estimated components, i.e., the regression coefficients, are decreased by using the threshold to throw away small regression coefficients. Accordingly, we list the biologically validated

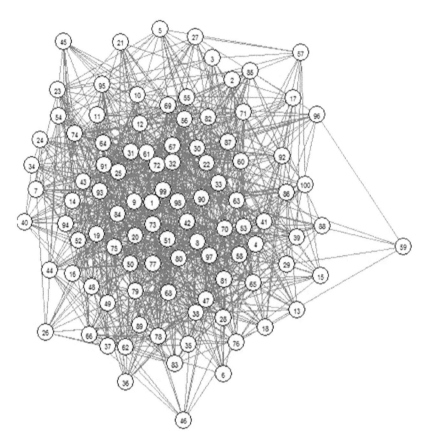

FIGURE 7.9
Estimated structure of the human gene interaction network via MARS with interaction effects.

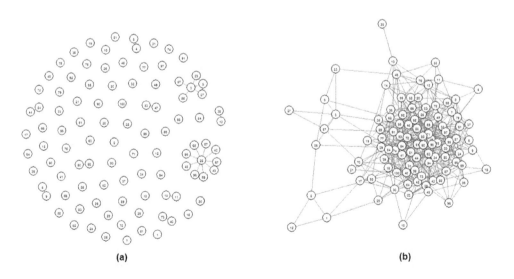

(a)   (b)

FIGURE 7.10
Estimated structure of the human gene interaction network via GGM with (a) CAIC (b) ICOMP and CAICF model selection criteria.

interactions of this dataset via both MARS models and GGM whose inference is conducted by the approximate glasso approach. From this assessment, we see that all three methods can detect 26 interactions, which are biologically supported by the study of Bhadra and Mallick [49] and also supported by STRING and GeneMANIA databases. But MARS with interactions can also find six more edges with respect to the findings via MARS without interaction. Finally, GGM under approximate glasso via ICOMP, CAIC, and CAICF model selection criteria failed to catch the links. Therefore, similar to previous analyses, we see that the application of GGM can be limited for small- and moderate-dimensional systems since its inference can be better performed via approximate methods for high-dimensional networks. On the other hand, MARS with an interaction model can more successfully estimate the true interactions than the estimation of the system via MARS without an interaction model (Table 7.7).

7.4 Conclusion

In this study, we have proposed two new non-parametric models for the representation of biological networks. In the first part, we have suggested the RFA as it does not depend on the selection of the threshold or penalty value, which directly controls the structure of the system. In the analyses via simulated data, we have generated datasets under the multivariate normality and various joint densities that have distinct marginal distributions connected by the Gaussian copula. From the results, we have observed that RFA can be more preferable for normal measurements, but for non-normal datasets, the accuracies decrease. Whereas, from the analyses via real datasets, we have seen that RFA produces promising outputs and that it can be used as an alternative approach for the description of the steady-state activation of biological networks.

In the second part of the study, we have implemented the MARS method with interaction terms as an alternative of GGM. The performance of this new model was evaluated via the same real datasets with biological interpretations of the results. From the outcomes of both datasets, we have concluded that if the interaction terms are included into the model, it can detect new links which have not been mostly estimated by other methods even though their findings are biologically declared. Therefore, we believe that this extended version of MARS can be a plausible alternative of GGM and can estimate the true structure of the system better than the lasso-based MARS without interaction effects. Furthermore, we have seen that the application of GGM can be limited for high-dimensional systems due to its calculation of the inference. We can perform some approximate methods under this condition, whereas neither MARS models has such a limitation and can be applicable for realistically complex biological networks.

As a future study, the PSICOV (precise structure contact prediction using sparse inverse covariance estimation) approach, which uses a sparse covariance estimation to the problem of protein contact prediction, can be adopted to infer the precision matrix [55]. Also, we can apply BIG and QUICK methods which include a block-coordinate descent method with the blocks via clustering [56]. Moreover, we suggest using alternative methods of MARS, known as POLYMARS [57], hybrid adaptive splines (HAS) [58], Bayesian MARS [59,60], SARS [61], and conicmultivariate adaptive regression splines (CMARS) [37,62] by converting them to a lasso regression in order to accelerate the speed of calculations and to increase the accuracies of the estimated systems since their advantages

TABLE 7.7

Comparison of the Estimated Links via MARS with/without Interaction Effects and GGM with Approximate Glasso

Links	MARS with Interaction	MARS without Interaction	GGM via CAIC	GGM via CAICF and ICOMP
Common genes	GI.27754767.I–GI.16554578.S	GI.27754767.I–GI.16554578.S	GI.27754767.I–GI.16554578.S	GI.27754767.I–GI.16554578.S
	GI.9961355.S–GI.27754767.I	GI.9961355.S–GI.27754767.I	GI.9961355.S–GI.27754767.I	GI.9961355.S–GI.27754767.I
	GI.27754767.I–GI.27754767.A	GI.27754767.I–GI.27754767.A	GI.27754767.I–GI.27754767.A	GI.27754767.I–GI.27754767.A
	GI.22027487.S–GI.27754767.I	GI.22027487.S–GI.27754767.I	GI.22027487.S–GI.27754767.I	GI.22027487.S–GI.27754767.I
	GI.20302136.S–GI.7661757.S	GI.20302136.S–GI.7661757.S	GI.20302136.S–GI.7661757.S	GI.20302136.S–GI.7661757.S
	GI.17981706.S–GI.13514808.S	GI.17981706.S–GI.13514808.S	GI.17981706.S–GI.13514808.S	GI.17981706.S–GI.13514808.S
Different genes	GI.7019408.S–GI.4504436.S	GI.7019408.S–GI.4504436.S		GI.7019408.S–GI.4504436.S
	GI.28610153.S–GI.4504436.S	GI.28610153.S–GI.4504436.S		GI.28610153.S–GI.4504436.S
	GI.20070269.S–GI.28610153.S	GI.20070269.S–GI.28610153.S		GI.20070269.S–GI.28610153.S
	GI.18379361.A–GI.20070269.S	GI.18379361.A–GI.20070269.S		GI.18379361.A–GI.20070269.S
	GI.4505888.A–GI.41350202.S	GI.4505888.A–GI.41350202.S		GI.4505888.A–GI.41350202.S
	GI.38569448.S–GI.22027487.S	GI.38569448.S–GI.22027487.S		GI.38569448.S–GI.22027487.S
	GI.34222299.S–GI.22027487.S	GI.34222299.S–GI.22027487.S		GI.34222299.S–GI.22027487.S
	GI.21614524.S–GI.34222299.S	GI.21614524.S–GI.34222299.S		GI.21614524.S–GI.34222299.S
	GI.37537705.I–GI.31652245.I	GI.37537705.I–GI.31652245.I		GI.37537705.I–GI.31652245.I
	GI.18641371.S–GI.41197088.S	GI.18641371.S–GI.41197088.S		GI.18641371.S–GI.41197088.S
	GI.16159362.S–GI.31652245.I	GI.16159362.S–GI.31652245.I		GI.16159362.S–GI.31652245.I
	GI.21389558.S–GI.16159362.S	GI.21389558.S–GI.16159362.S		GI.21389558.S–GI.16159362.S
	GI.28557780.S–GI.16159362.S	GI.28557780.S–GI.16159362.S		GI.28557780.S–GI.16159362.S

(Continued)

TABLE 7.7 (*Continued*)

Comparison of the Estimated Links via MARS with/without Interaction Effects and GGM with Approximate Glasso

Links	MARS with Interaction	MARS without Interaction	GGM via CAIC	GGM via CAICF and ICOMP
	GI.27477086.S − GI.16159362.S	GI.27477086.S − GI.16159362.S		GI.27477086.S − GI.16159362.S
	GI.23510363.A − GI.28557780.S	GI.23510363.A − GI.28557780.S		GI.23510363.A − GI.28557780.S
	GI.27482629.S − GI.23510363.A	GI.27482629.S − GI.23510363.A		GI.27482629.S − GI.23510363.A
	GI.28416938.S − GI.27482629.S	GI.28416938.S − GI.27482629.S		GI.28416938.S − GI.27482629.S
	GI.30795192.A − GI.27482629.S	GI.30795192.A − GI.27482629.S		GI.30795192.A − GI.27482629.S
	GI.24308084.S − GI.27477086.S	GI.24308084.S − GI.27477086.S		GI.24308084.S − GI.27477086.S
	GI.4504700.S − GI.19224662.S	GI.4504700.S − GI.19224662.S		GI.4504700.S − GI.19224662.S
	GI.14211892.S − GI.20373176.S			
	GI.5454143.S − GI.4504410.S			
	GI.33356162.S − GI.17981706.S		GI.33356162.S − GI.17981706.S	GI.33356162.S − GI.17981706.S
	GI.17981706.S − GI.14211892.S		GI.17981706.S − GI.14211892.S	GI.17981706.S − GI.14211892.S
	GI.27894333.A − GI.27477086.S		GI.27894333.A − GI.27477086.S	GI.27894333.A − GI.27477086.S
	GI.20373176.S − GI.14211892.S		GI.20373176.S − GI.14211892.S	GI.20373176.S − GI.14211892.S
			GI.17981706.S − GI.33356559.S	GI.17981706.S − GI.33356559.S
			GI.20070269.S − GI.28610153.S	GI.20070269.S − GI.28610153.S
			GI.33356559.S − GI.17981706	GI.33356559.S − GI.17981706
Total links:	32	26	13	33

Source: The true links are taken from the study of Bhadra and Mallick [49].

over the full description of MARS have been indicated. Finally, we aim to evaluate the performance of both MARS models and GGM in different types of networks and various distributions of observations.

Acknowledgments

The authors thank Prof. Dr. Hamparsum Bozdoğan and the anonymous referee for their valuable comments, which improved the quality and readability of this work. Moreover, they thank the Scientific and Technological Research Council of Turkey (TÜBİTAK) grant (Project no:114E636) and the BAP project at the Middle East Technical University (Project no: BAP-01-09-2016-002) for their support.

References

1. Wilkinson, D.J. (2011), *Stochastic Modelling for Systems Biology.* CRC press. https://doi.org/10.1201/9781351000918
2. Bower, J. M. and Bolouri, H. (2001), *Computational Modeling of Genetic and Biochemical Networks.* MIT press. ISBN:9780262524230, 0262524236
3. Jong, D.H. (2002), Modeling and simulation of genetic regulatory systems: a literature review, *Journal of Computational Biology*, 9 (1), 67–103.
4. Golightly, A. and Wilkinson, D. J. (2005), Bayesian inference for stochastic kinetic models using a diffusion approximation, *Biometrics*, 61 (3), 781–788.
5. Purutçuoğlu, V. (2013), Inference of the stochastic MAPK pathway by modified diffusion bridge method, *Central European Journal of Operations Research*, 1–15.
6. Whittaker, J. (1990), *Graphical Models in Applied Multivariate Statistics*, New York: John Wiley and Sons.
7. Meinshausen, N. and Buhlmann, P. (2006), High-dimensional graphs and variable selection with the lasso, *The Annals of Statistics*, 34, 1436–1462.
8. Yuan, M. and Lin, Y. (2007), Model selection and estimation in the Gaussian graphical model, *Biometrika*, 94, 19–35.
9. Banerjee, O., Ghaoui, E.L., and d'Aspremont, A. (2008), Model selection through sparse maximum likelihood estimation for multivariate Gaussian or binary data, *Journal of Machine Learning Research*, 9, 485–516.
10. Drton, M. and Perlman, M.D. (2008), A SINful approach to Gaussian graphical model selection, *Journal of Statistical Planning and Inference*, 138 (4), 1179–1200.
11. Friedman J., Hastie T., and Tibshirani R. (2007), Pathwise Coordinate Optimization. *The Annals of Applied Statistics*, 1 (2), 302–332.
12. Tibshirani R. and Saunders M. (2005), Sparsity and smoothness via the fused lasso, *Journal of the Royal Statistical Society*, 67 (1), 91–108.
13. Zou, H. (2006), The adaptive lasso and its oracle properties, *Journal of American Statistical Association*, 101, 1418–1429.
14. Zou, H.and Hastie, T. (2005), Regularisation and variable selection via the elastic net, *Journal of the Royal Statistical Society*, 67 (2), 301–320.
15. Ravikumar, P., Wainwright, M.J. and Lafferty, J.D. (2010), High-dimensional ising model selection using l_1-regularized logistic regression, *The Annals of Statistics*, 38(3), 1287–1319.

16. Nelsen, R. B. (2007), *An Introduction to Copulas*. Springer Science and Business Media. ISBN:9780387286785, 0387286780

17. Dobra, A. and Lenkoski, A. (2011), Copula Gaussian graphical models and their application to modeling functional disability data, *The Annals of Applied Statistics*, 5(2A), 969–993.

18. Dobra, A., Lenkoski, A. and Rodriguez, A. (2011), Bayesian inference for general Gaussian graphical models with application to multivariate lattice data, *Journal of the American Statistical Association*, 106 (496), 1418–1433.

19. Wang, H., Li, S.Z., et al. (2012), Efficient Gaussian graphical model determination under GWishart prior distributions, *Electronic Journal of Statistics*, 6, 168–198

20. Mohammadi, A. and Wit, E. (2015), BD graph: an R package for Bayesian structure learning in graphical models, *Bayesian Analysis*, 10 (1), 109–138.

21. Ihmels, J., Friedlander, G., Bergmann, S., Sarig, O., Ziv, Y. and Barkai, N. (2002), Revealing modular organization in the yeast transcriptional network, *Nature genetics*, 31 (4), 370–377.

22. Ayyıldız, E., Ağraz, M. and Purutçuoğlu, V. (2017), MARS as an alternative approach of Gaussian graphical model for biochemical networks, *Journal of Applied Statistics*, 44c(16), 2858–2876.

23. Breiman L. (2001), Random forest. *Machine Learning*, 45, 5–32.

24. Lewis, R. J. (2000, May). An introduction to classification and regression tree (CART) analysis. *Annual meeting of the society for academic emergency medicine in San Francisco, California*, Vol. 14.

25. Timofeev, R. (2004), Classification and regression trees (CART) Theory and Applications. Center of Applied Statistics and Economics Humboldt University, Berlin. (Master's Thesis).

26. Lewis, P.A.W. and Stevens, J.G.(1991), Nonlinear modeling of time series using multivariate adaptive regression splines, *Journal of American Statistical Association*, 86 (416), 864–877.

27. Haas, H. and Kubin, G. (1998), A multi-band nonlinear oscillator model for speech, *Conference Record of the Thirty-Second Asilomar Conference on Signals, Systems and Computers*, 338–342.

28. Chou, S.M., Lee, T.S., Shao, Y.E. and Chen, I.F. (2004), Mining the breast cancer pattern using artificial neural networks and multivariate adaptive regression splines, *Expert Systems with Applications*, 27, 133–142.

29. Ağraz, M. and Purutçuoğlu, V. (2016), Different types of Bernstein operators in inference of Gaussian graphical model, *Cogent Mathematics*, 3(1), 1–11.

30. Ağraz, M., and Purutçuoğlu, V. (2016). Transformations of data in deterministic modelling of biological networks, 343–356, in *Intelligent Mathematics II: Applied Mathematics and Approximation Theory*, Cham: Springer.

31. Hastie T. (2001), *The Elements of Statistical Learning*. New York: Springer, Verlag.

32. Lauritzen, S.L. (1996), *Graphical Models*, New York: Oxford University Press.

33. Tibshirani, R. (1996), Regression shrinkage and selection via the lasso, *Journal of the Royal Statistical Society*, 58 (1) 267–288.

34. Wit, E., Vinciotti, V. and Purutçuoğlu, V. (2010), Statistics for biological networks: short course notes. *25th International Biometric Conference (IBC)*, Florianopolis, Brazil.

35. Friedman, J., Hastie, T., and Tibshirani, R. (2008), Sparse inverse covariance estimation with the graphical lasso, *Biostatistics*, 9, 432–441.

36. Friedman, J. (1991), Multivariate adaptive regression splines, *The Annual of Statistics*, 19 (1), 1–67.

37. Batmaz, I, Yerlikaya, F., Kartal, E., Koksal, G. and Weber, G.W. (2010), Evaluating the CMARS performance for modeling nonlinearities. Power control and optimization. *Proceedings of the 3rd Global Conference on Power Control and Optimization, Gold Coast*, Australia, 2–4 February, 351–357.

38. Craven, P. and Wahba, G. (1979), Smoothing noisy data with spline functions, *Numerical Mathematics*, 31, 377–403.

39. Kartal-Koc, E. and Bozdogan, H. (2015), Model selection in multivariate adaptive regression splines (MARS) using information complexity as the fitness function, *Machine Learning*, 101 (1–3), 35–58.

40. Bozdogan, H. (1988), ICOMP: A new model selection criterion, 599–608 in *Classification and Related Methods of Data Analysis*, Ed. Hans H. Bock, Amsterdam: Elsevier Science Publishers B. V. (North-Holland).

41. Bozdogan H. (1990), On the information-based measure of covariance complexity and its application to the evaluation of multivariate linear models, *Communications in Statistics Theory and Methods*, 19(1), 221–278.

42. Bozdogan H. (1987), Model selection and Akaike's information criterion (AIC): the general theory and its analytical extensions, *Psychometrika*, 52 (3), 345–370.

43. Alon, U. (2006), *An Introduction to Systems Biology: Design Principles of Biological Circuits*, Chapman and Hall/CRC, https://doi.org/10.1201/9781420011432

44. Barabasi A.L. and Oltvai Z.N. (2004), Network biology: understanding the cell's functional organization, *Nature Review Genetics*, 5, 2101–2113.

45. Liu, X., Milo, M., Lawrence, N. and Rattray, M. (2005), A tractable probabilistic model for Affymetrix probe-level analysis across multiple chips, *Bioinformatics*, 21 (18), 3637–3644.

46. Finner, H., Dickhaus, T. and Roters, M. (2008), Asymptotic tail properties of Student's t distribution, *Communications in Statistics Theory and Methods*, 37 (2), 175–179.

47. Sachs, K., Perez, O., Pe'er, D., Lauffenburger, D.A. and Nolan, G.P. (2005), Causal protein signaling networks derived from multiparameter single-cell data, *Science*, 308 (5721), 523– 529.

48. Stranger, B., Nica, A., Forrest, M., Dimas, A., Bird, C., Beazley, C., Ingle, C., Dunning, M., Flicek, P., Montgomery, S., Tavare, S., Deloukas, P. and Dermitzakis, E. (2007), Population genomics of human gene expression, *Nature Genetics*, 39, 1217–1224.

49. Bhadra A. and Mallick B.K. (2013), Joint high-dimensional Bayesian variable and covariance selection with an application to eQTL analysis, *Biometrics*, 69 (2), 447–457.

50. Chen, L., Emmert-Streib, F. and Storey, J. (2007), Harnessing naturally randomized transcription to infer regulatory relationships among genes, *Genome Biology*, 8, R219.

51. Friedman, J., Hastie, T., and Tibshirani, R. (2014), Glasso: Graphical lasso estimation of Gaussian graphical models (R package).http://www-stat.stanford.edu/~tibs/glasso

52. Sim, A.T.R and Scott, J.D. (1999), Targeting of PKA, PKC and protein phosphatases to cellular microdomains, *Cell Calcium*, 26 (5), 209–217.

53. Cunningham, J., Estrella, V., Lloyd, M., Gillies, R.,Frieden, B.R. and Gatenby, R. (2012), Intracellular electric field and pH optimize protein localization and movement, *PLoS One*, 7, 1–12.

54. Kuo, J.F. (1994), *Protein Kinase C*, New York: Oxford University Press.

55. Jones, D., Buchan, D.W.A., Cozzetto, D. and Pontil, M. (2012), PSICOV: precise structural contact prediction using sparse inverse covariance estimation on large multiple sequence alignments, *Bioinformatics*, 28 (2), 184–190.

56. Hsieh, C.J, Sustik, M., Dhillon, I.S., Ravikumar, P. and Poldrak, R.A. (2013), Advances in Neural Information Processing Systems 26. *29th Annual Conference on Neural Information Processing Systems 2015*, 3165–3173.

57. Stone, C., Hansen, M., Kooperberg, C. and Troung, Y.(1997), Polynomial splines and their tensor products in extended linear modeling, *Annals of Statistics*, 25, 1371–1470.

58. Lou, Z. and Wahba, G. (1997), Hybrid adaptive splines, *Journal of the American Statistical Association*, 92, 107–116.

59. Denison, D.G.T, Mallick, B.K. and Smith, A.F.M. (1998), Automatic Bayesian curve fitting, *Journal of Royal Statistical Society*, 60, 333–350.

60. DiMatteo, I., Genovese, C. R. and Kass, R. E. (2001), Bayesian curve fitting with free-knot splines, *Biometrika*, 88, 1055–1071.

61. Zhou, S. and Shen, X. (2012), Spatially adaptive regression splines and accurate knot selection schemes, *Journal of American Statistical Association*, 96, 247–259.

62. Yerlikaya, F. (2008), A new contribution to nonlinear robust regression and classification with MARS and its application to data mining for quality control in manufacturing. MSc Thesis, Graduate School of Applied Mathematics, Middle East Technical University, Ankara, Turkey.

8

Generating Various Types of Graphical Models via MARS

E. Ayyıldız
Turkish National Research Institute of Electronics and Cryptology (TÜBİTAK BİLGEM)

V. Purutçuoğlu
Middle East Technical University

CONTENTS

8.1 Introduction

There are several modelling approaches to describe the structure of complex biological systems. The Gaussian graphical model (GGM) is one of the well-known mathematical models to estimate the undirected structure of the biological networks [1]. In GGM, the conditional independency is used to construct the graphical model under the multivariate normal distribution. In this model, the independency is denoted by the 0 entry in the precision matrix, which is the inverse of the covariance matrix under the lasso regression [2,3]. In this regression, each node is regressed against all remaining nodes when the network is sparse. In inference of this model, various methods have been proposed. For instance, Li and Gui (2006) [4] suggest the threshold gradient descent approach and Friedman et al. (2008) [5] propose a block-wise coordinate descent method to estimate the precision matrix. But the computational demand of these methods is challenging, especially, for large systems.

On the other hand, the multivariate adaptive regression spline (MARS) is a widely used nonparametric regression method that can be applicable for high-dimensional data under nonlinearity [6]. Previous studies have showed that MARS can be an alternative to GGM if it is estimated by the main effects when all the interactions and nonlinear components are eliminated from the model [7]. In this lasso-based MARS model, the pairwise links between the species of the system are determined by important terms of the estimated model. From the analyses, it has been seen that the MARS model is as accurate as GGM with a significant gain in the computational demand.

Therefore, in this study, after obtaining the coefficients of the underlying lasso-based MARS model, we suggest to proportionally convert these coefficients to the Pearson,

Spearman, and Kendall's tau correlation coefficients as a filter in order to eliminate the overestimated links, i.e., to control the false positive rate, in the estimated networks. For the comparison of the performance, we calculate different accuracy measures such as the specificity, F-measure, and Matthews correlation coefficient (MCC) based on the Monte Carlo runs under distinct dimensional datasets for all proposed methods. Also, we conduct simulation studies under both normal and non-normal data in order to evaluate the robustness of the results. Moreover, a real biochemical dataset is applied to assess the performance of the suggested methods. Finally, we summarize the results and propose future work. Accordingly, in the organization of the study, we present the suggested approaches in Section 8.2. In Section 8.3, we list the application and the outcomes of the comparative analyses. We conclude the findings in Section 8.4.

8.2 Lasso-Based MARS and the Relation with GGM

The GGM is one of the most widely used parametric approaches to construct the undirected graphical model of the biochemical networks. GGM makes some basic assumptions. These can be listed as a mean term with 0 value, a variance–covariance matrix shown by Σ, and a multivariate normal distribution for observations. In GGM, the absence of the edge between two nodes corresponds to the conditional independency of these nodes. This information is also represented by the 0 entry in the inverse of Σ. This special inverse matrix is called the precision matrix. The conditional independent structure of nodes is defined by the regression model in such a way that a model is constructed for each node against all remaining nodes. Hence, the 0 coefficient of this regression model, which is called the lasso model, indicates the conditional independency of the corresponding nodes by the following expression.

$$Y^{(-p)} = \beta \, Y^{(-p)} + \varepsilon \tag{8.1}$$

where $Y(p)$ and $Y(-p)$ describe the state of the pth node and the states of all other nodes except the node p, respectively. Moreover, ε is the independent and normally distributed error term with a 0 mean and a constant variance. Furthermore, there is a relation between the coefficients β and the precision matrix Θ [8] in this model via

$$\beta = -\Theta_{-p,p} / \Theta_{p,p} \tag{8.2}$$

In Equation 8.2, $\Theta_{p,p}$ shows the pth diagonal entry of Θ, and $\Theta_{-p,p}$ stands for the off-diagonal entry of Θ when the pth node is excluded.

In GGM, to estimate the precision matrix, a widely used method is the maximization of the penalized log-likelihood function via the graphical lasso (glasso) [5]. The penalized log-likelihood function, which is maximized over nonnegative defined matrices, is presented as follows:

$$\max_{\Theta} \left[\log(|\Theta|) - \mathrm{Tr}(S\Theta) - \lambda \, \|\Theta\|_1 \right] \tag{8.3}$$

in which S denotes the sample covariance matrix, Tr(.) represents the trace, and $||.||1$ is the $l1$-norm, which is the sum of the absolute value of a given element. Additionally, λ shows the nonnegative penalizing constant and it has a crucial role in the construction of the optimal model. Because it controls the sparsity of the estimated precision matrix in such a way, a large value of λ implies very sparse precision matrix Θ, whereas a small λ presents non-sparse Θ. Here, one of the widely used approaches to find the optimal λ is the k-cross validation method [9].

On the other hand, the multivariate adaptive regression spline (MARS) is a very effective nonparametric regression approach due to the fact that it can successfully deal with high-dimensional and correlated data that have a nonlinear structure [6]. Basically, this model is a specific regression model designed for multivariate and univariate data. Hence, in modelling, MARS can generate a hierarchical model that is composed of basis functions. The underlying basis functions are found by a stepwise selection [10]. This approach builds a model that includes main and interaction terms. Also, it is possible to restrict it to the order of interactions for the candidate terms with a specified limit [10]. Moreover, the complexity in nonlinear functions can be reduced by constructing only the linear models. In general, the MARS model consists of a two steps: forward step and backward step. In the first step, the model is initialized via an intercept term. Later, the model is extended iteratively via the basis function. The underlying functions are chosen by using the observations by means of a stepwise calculation [10]. Finally, a complex model that contains many basis functions is obtained. But this model can have the problem of overfitting. Hereby, in each backward stage, the term which causes the smallest increase in the residual squared error is removed from the model to prevent the underlying challenge.

MARS uses the piecewise linear basis functions $(x-t)_+$ and $(t-x)_+$, which are described as follows:

$$(x - t)_+ = \begin{cases} (x-t) & \text{if } x > t, \\ 0 & \text{otherwise} \end{cases} \quad \text{and} \quad (t - x)_+ = \begin{cases} (t-x) & \text{if } x > t, \\ 0 & \text{otherwise.} \end{cases} \tag{8.4}$$

Here, the (+) sign represents the positive part. These piecewise linear functions are linked to each other at the value t, called the knot. Furthermore, the associated basis functions are named as linear splines. In the end, x_j denoting the independent variable is represented by a pair of expressions including basis functions as below:

$$C = \left\{ (x_j - t)_+, (t - x_j)_+ \right\}, \tag{8.5}$$

where $t \in \{x_{1\,j}, x_{2\,j}, ..., x_{n-j}\}$ ($j = 1, 2, ..., p$) and p is the number of independent variables.

In MARS, the functions from the set of basis functions C are used for the model building instead of the original independent variables. Hence, the model can be described as

$$f(x) = \beta_0 + \sum_{m=1}^{M} \beta_m h_m(x) + \varepsilon \tag{8.6}$$

in which $h_m(x)$ is a function in the set C, and M shows the total number of basis functions in the model. Hence, the coefficients β_m are inferred via optimization in such a way that the residual sum of squares is minimized. These coefficients can be seen as weights that define the importance of the variables. This model is constructed at the end of the forward stage.

Then, in order to estimate the optimal, i.e., best, model which includes λ amount of terms, the model eliminates some terms via the backward stage. In this searching process, the optimal λ is defined via a model selection criterion, named the *generalized cross-validation* (GCV) value defined in its explicit form as

$$\mathrm{GCV}(\lambda) = \frac{\sum_{i=1}^{N} \left(y_i - \hat{f}_\lambda(x_i)\right)^2}{\left(1 - M(\lambda)/N\right)^2}, \tag{8.7}$$

where N denotes the number of observations and $M(\lambda)$ represents the effective number of parameters, which is formulated as $M(\lambda) = r + cK$. Additionally, in the above expression, r implies the number of basis functions that should be linearly independent. K represents the total number of knot points that are specifically applied in the forward step of the model. Moreover, the cost of the optimization while computing the basis function is shown by c and it is typically equated to three depending on the structure of the model. That is, if there is a constraint about the additive structure of the model, then this c number is set to 2. On the other hand, the above term in Equation 8.6 denotes the residual sum of squares. Lastly, $\hat{f}_\lambda(x_i)$ in the numerator of the same equation indicates the estimated model, which is composed of λ amount of terms.

From previous studies, it has been shown that if MARS is estimated by discarding all interactions and nonlinear components, resulting in a sole inference of the main effects, it can be a strong alternative of GGM since the final model is constructed as in Equation 8.1 [7]. We call this type of summarized MARS model the *lasso-based MARS model*, which is also computationally more efficient than GGM. Furthermore, in the construction of the network via the optimal lasso-based MARS model, it is assumed that the estimated regression coefficients, i.e., the *important terms* of the estimated model, can be the indication of the pairwise links between species in the system.

In this study, after obtaining the important coefficients of the lasso-based MARS model, we proportionally convert these coefficients to the *Spearman* and *Kendall's tau correlation coefficients*. The underlying transformation is not designed as one-to-one since we define the variance terms, which are the diagonals of the precision matrix, as 1. The reason is that the nodes in the biological systems, modeled via GGM, are assumed to be autoregulated [1]. In the calculation, we use the following equations to convert the proportional Pearson correlation matrix found from β as in Equations 8.1 and 8.2 to Spearman and Kendall's tau correlation coefficients, respectively [11,12]:

$$\rho_S = \frac{6}{\pi} \arcsin\left(\frac{\rho}{2}\right), \tag{8.8}$$

$$\tau = \frac{2}{\pi} \arcsin(\rho), \tag{8.9}$$

where ρ represents the Pearson correlation coefficient, ρ_S denotes the Spearman, and τ refers to Kendall's tau values. In all these calculations, we consider that the selected non-parametric correlation coefficients can better represent the feature of the nonparametric models such as MARS. On the other side, as the relation between β and Θ in Equation 8.2 is valid for the Pearson correlation coefficient, we compute its correspondence under ρ_S and τ.

Finally, to compare the performance of the lasso-based MARS model under the current selection procedure and newly suggested filters, i.e., distinct choices of correlation coefficients from Pearson, Spearman, and Kendall's tau, we compute their *specificities, F-measures,* and MCCs, as presented below [13]:

$$\text{Specificity} = \text{TN}/(\text{TN+FP}), \tag{8.10}$$

$$\text{F-measure} = 2\frac{\text{Precision} \times \text{Recall}}{\text{Precision} + \text{Recall}}, \tag{8.11}$$

$$\text{MCC} = \frac{\text{TP} \times \text{TN} - \text{FP} \times \text{FN}}{\sqrt{(\text{TP} + \text{FP})(\text{TP} + \text{FN})(\text{TN} + \text{FP})(\text{TN} + \text{FN})}}, \tag{8.12}$$

where

$$\text{Precision} = \text{TP}/(\text{TP+FP}), \tag{8.13}$$

$$\text{Recall} = \text{TP}/(\text{TP+FN}), \tag{8.14}$$

In these equations, the true positive, the true negative, false positive, and false negative are shown by TP, TN, FP, and PN, respectively. As used in many studies, TP presents the number of correctly classified objects with positive labels and TN the number of correctly classified objects with negative labels. Similarly, FP and FN refer to the number of misclassified objects with negative and positive labels, in that order.

8.3 Applications

In the application of MARS, we use both simulated and real datasets. In the simulation part, we generate datasets from the multivariate normal distribution by using the *huge* package in R [14]. The datasets are generated under various dimensions, namely, (50×50), (100×100), and (300×300), and the number of observations for each data is set to 20, arbitrarily. Moreover, since the biological networks, such as the protein regularity and the cellular metabolism networks, have a scale-free feature [15], we also produce datasets under this condition. The estimated coefficients are found via the lasso-based MARS model and the different correlation values are computed from the corresponding models. In the calculation via MARS, we consider that there is an interaction, i.e., edge, between two nodes if there is a regression coefficient for each constructed gene-specific model as in Equation 8.1. Furthermore, in inference of the network, we accept that if either the (i, j) or (j, i) entry of the coefficient matrix has a value, rather than the 0 entry, we can assign 1 for both the (i, j) and (j, i) entries of the estimated adjacency matrix. On the other hand, we set the threshold value to 0.6 as an indication of the significant correlation coefficient to convert the estimated ρ to the binary adjacency matrix. Thus, the coefficients greater than 0.6 are taken as 1, otherwise, they are equated to 0 in the final structure of the system. But we use this threshold when we generate data from the multivariate normal distribution. Whereas for the non-normal

TABLE 8.1

Comparison of the Specificity, F-Measure and MCC for Systems with 50, 100 and 300 Genes via Lasso-Based MARS under Important $\hat{\beta}$ (Important $\hat{\beta}$), Estimated Pearson Correlation Coefficient (Pearson) Matrix, Spearman Correlation Coefficient Matrix (Spearman) and Kendall's Tau Correlation Coefficient Matrix (Kendall)

Total Number of Genes	Method	Specificity	F-Measure	MCC
50	Important $\hat{\beta}$	0.8957	0.3016	0.2612
	Pearson	0.9911	0.4723	0.4871
	Spearman	0.9920	0.4752	0.4935
	Kendall	0.9974	0.4947	0.5402
100	Important $\hat{\beta}$	0.9387	0.2397	0.2285
	Pearson	0.9972	0.4757	0.5126
	Spearman	0.9976	0.4786	0.5184
	Kendall	0.9994	0.4958	0.5562
300	Important $\hat{\beta}$	0.9769	0.1956	0.2069
	Pearson	0.9997	0.4863	0.5442
	Spearman	0.9998	0.4881	0.5480
	Kendall	1.0000	0.4976	0.5687

datasets, we change this value with respect to the mean estimated ρ so that there is no bias in the calculation. Finally, we compute the selected accuracy measures to compare the performance of all proposed methods. The results based on the 1,000 Monte Carlo runs for each dimension are presented in Table 8.1. From the findings, we see that Pearson, Spearman, and the Kendall's tau correlation coefficients have greater specificity, F-measure, and MCC values regarding the lasso-based MARS model found via estimated β, i.e., $\hat{\beta}$ for each dimensional matrix. Here, Pearson's and Spearman's results are very close to each other and Kendall's tau has greater values for every accuracy measure. Furthermore, the specificity values for all methods are relatively close to each other. However, they indicate differences in terms of the F-measure and MCC values. Additionally, when the dimension of matrices increases, the F-measure and MCC values of the MARS method decrease. But the corresponding values of other alternatives increase. As a result, we show that the construction of the lasso-based MARS model improves the accuracy of the estimated systems if it is based on the correlation coefficients, rather than the entry of the β regression coefficients.

Then, in order to evaluate the performance of the suggested approaches, in non-normal datasets, we generate non-normal joint distributions via the *Gaussian copula* whose independent marginal distributions are exponential with the rate 4. Furthermore, to get comparable results with respect to the real data, we perform the Monte Carlo iterations under the same and also lower-dimensional systems. Therefore, the dimensions of the datasets are taken as 10, 50, and 100 nodes and the number of observations for each scale-free data is set to 20, as used in the previous scenarios. Then, we apply the same steps of calculations to convert the estimated β's to the different correlation matrices, resulting in their corresponding adjacency matrices. Here, as the threshold value in each binary adjacency matrix we choose the mean entry of each associated estimated correlation matrix rather than a fixes value of 0.6 as used in multivariate normal data. Thereby, the entries lower than the mean of the entries in the correlation matrix are set to 0, otherwise, they are equated to 1.

TABLE 8.2

Comparison of the Specificity, F-Measure and MCC for 10, 50, and 100 Dimensional Systems Whose Data Are Generated via the Gaussian Copula with Exponential Marginals and Whose Estimated Adjacency Matrices Are Computed by the Lasso-Based MARS with Important $\hat{\beta}$ (Important $\hat{\beta}$), and Proportional to the Pearson Correlation Coefficient (Pearson), Spearman Correlation Coefficient (Spearman) as well as Kendall's Tau Correlation Coefficient (Kendall)

Total Number of Genes	Method	Specificity	F-Measure	MCC
50	Important $\hat{\beta}$	0.7098	0.4754	0.2397
	Pearson	0.8257	0.4904	0.3065
	Spearman	0.8344	0.4925	0.3139
	Kendall	0.9206	0.5074	0.3960
100	Important $\hat{\beta}$	0.8788	0.2493	0.2016
	Pearson	0.9472	0.3361	0.2918
	Spearman	0.9506	0.3427	0.2993
	Kendall	0.9770	0.4102	0.3860
300	Important $\hat{\beta}$	0.9353	0.2174	0.2039
	Pearson	0.9730	0.3167	0.2953
	Spearman	0.9746	0.3238	0.3025
	Kendall	0.9881	0.3979	0.3878

In this assessment, we also control $\mu \mp 2\sigma$ and $\mu \mp 3\sigma$, where σ refers to the standard deviation in the entries of the estimated correlation coefficient, to decide on the cutoff values. But since we observe no difference in the transformation of the adjacency matrix under these conditions either, we take the mean values for simplicity. All outcomes are presented in Table 8.2. Similar to previous analyses, Pearson, Spearman, and Kendall's tau correlation coefficients indicate higher specificity, F-measure, and MCC values with respect to the lasso-based MARS model under the important regression coefficients. Moreover, among alternatives, Kendall's tau has the best accuracy value. Furthermore, when we increase the dimension of the system, the specificity values become higher, whereas F-measure and MCC get smaller.

On the other hand, in order to evaluate the performance of the suggested methods in real biochemical data, we use the single cell dataset in the study of Sachs et al. (2005) [16]. To construct the underlying cellular protein-signaling network, we apply the simultaneous measurements of the multiple phosphorylated proteins and phospholipid components in thousands of individual primary human immune system cells. The aim of this analysis is to understand the native-state tissue signaling biology, complex drug actions, and the dysfunctional signals in diseased cells [16]. Hereby, the dataset contains the flow cytometry of 11 proteins measured on 11,672 red blood cells. From the analysis, it is observed that the specificity value of the lasso-based MARS model that depends on the important entry of the estimated β is less than in the other methods. Furthermore, the MCC values of Pearson, Spearman, and Kendall's tau correlation coefficients are greater than the lasso-based MARS method with significant $\hat{\beta}$. On the other hand, the value of F-measure via significant $\hat{\beta}$ and the Pearson correlation coefficient are very close to each other. These results are also in agreement with the non-normal simulated data analyses. In Table 8.3, we represent the associated values.

TABLE 8.3

Comparison of the Specificity, F-Measure and MCC for the Cell Signal Data via Lasso-Based MARS under Important $\hat{\beta}$ (Important $\hat{\beta}$), Estimated Pearson Correlation Coefficient (Pearson) Matrix, Spearman Correlation Coefficient Matrix (Spearman) and Kendall's Tau Correlation Coefficient Matrix (Kendall)

Method	Specificity	F-Measure	MCC
Important $\hat{\beta}$	0.7098	0.4754	0.2397
Pearson	0.8257	0.4904	0.3065
Spearman	0.8344	0.4925	0.3139
Kendall	0.9206	0.5074	0.3960

8.4 Conclusion

In this study, we have suggested alternative methods to increase the accuracy of the lasso-based MARS model used in the construction of biological networks. In previous analyses, it has been shown that if the important regression coefficients of the lasso-based MARS are directly taken for the description of the adjacency matrix (by setting unimportant ones to 0 and important ones to 1), it can be a promising alternative of the Gaussian graphical model. Here, we have implemented the Pearson, Spearman, and Kendall's tau correlation coefficients computed from the underlying estimated regression coefficients and converted them into estimated adjacency matrices via threshold values. From the comparative analyses based on the specificity, F-measure, and MCC in both simulated and real datasets, we have found that the newly proposed approaches improve most of the selected accuracy measures. Furthermore, these outputs have been detected under different dimensional systems and observations from both multivariate normal and non-normal datasets. Therefore, we have concluded that the lasso-based MARS is a strong alternative of the Gaussian graphical model and the true network can be better estimated by the correlation coefficient matrices, in particular, by Kendall's tau correlation matrix for different types of biological systems. As an extension of this study, we suggest other nonparametric models such as CMARS and RCMARS as the competitive models for both GGM and lasso-based MARS. Furthermore, we aim to analyze, specifically, the performance of the nonparametric correlation matrices in the estimation of true networks in such novel models. Moreover, we can still improve the filtering stage of the precision matrix by the weighted version of Kendall's tau [17] and Spearman coefficients [17,18].

References

1. Whittaker, J. (1990). *Graphical models in applied multivariate statistics.* John Wiley and Sons, New York.
2. Tibshirani, R. (1996). Regression shrinkage and selection via the lasso. *Journal of the Royal Statistical Society,* 58, (1) 267–288.
3. Meinshausen, N. and Bhlmann, P. (2006). High dimensional graphs and variable selection with the lasso. *The Annals of Statistics,* 34, 1436–1462.

4. Li, H., and Gui, J. (2006). Gradient directed regularization for sparse Gaussian concentration graphs with applications to inference of genetic networks. *Biostatistics*, 7, (2), 302–317.
5. Friedman, J., Hastie, T. and Tibshirani, R. (2008). Sparse inverse covariance estimation with the graphical lasso. *Biostatistics*, 9, 432–441.
6. Friedman, J.H. (1991). Multivariate adaptive regression splines. *The Annual of Statistics*, 19, (1), 1–67.
7. Ayyıldız, E., Ağraz, M. and Purutçuoğlu, V. (2017). MARS as an alternative approach of Gaussian graphical model for biochemical networks. *Journal of Applied Statistics*, 44, (16), 2858–2876.
8. Wit, E., Vinciotti, V., and Purutçuoğlu, V. (2010). Statistics for biological networks: short course notes. *25th International Biometric Conference (IBC)*, Florianopolis, Brazil.
9. Hastie, T., Tibshirani, R., Friedman, J. H. (2001). *The element of statistical learning*. Springer, Verlag, New York.
10. Barron, A. R., and Xiao, X. (1991). Discussion: multivariate adaptive regression splines. *The Annals of Statistics*, 19, (1), 67–82.
11. Genest, C. and Favre, A. N. (2007). Everything you always wanted to know about copula modeling but were afraid to ask. *Journal of Hydrologic Engineering*, 12, (4), 347–368.
12. Gibbons, J. D., and Chakraborti, S. (2003). *Nonparametric statistical inference*. Marcel Dekker, Inc., New York.
13. Fan, J., Feng, Y., and Wu, Y. (2009). Network exploration via the adaptive lasso and SCAD penalties. *The Annuals of Applied Statistics*, 3, (2), 521–541.
14. Zhou, S., Rtimann, P., Xu, M., and Bhlmann, P. (2011). High dimensional covariance estimation based on Gaussian graphical models. *Journal of Machine Learning Research*, 12, (4), 2975–3026.
15. Barabasi, A. L., and Oltvai, Z. N. (2004). Network biology: understanding the cell's functional organization. *Nature Reviews Genetics*, 5, 101–113.
16. Sachs, K., Perez, O., Pe'er, D., Lauffenburger, D. and Nolan, G. (2005). Causal protein-signaling networks derived from multiparameter single-cell data. *Science*, 308, (5721), 523– 529.
17. Kumar, R., and Vassilvitskii, S. (2010). Generalized distances between rankings. In *Proceedings of the 19th International Conference on World Wide Web, ACM*, 571–580.
18. Chicco, D., Ciceri, E., and Masseroli, M. (2015). Extended Spearman and Kendall coefficients for gene annotation list correlation. *CIBB*, 2014, 19–32.

9

Financial Applications of Gaussian Processes and Bayesian Optimization

Syed Hasan Jafar
Woxsen University

CONTENTS

9.1 Introduction

Gaussian cycles and Bayesian optimization (BO) are effective and sound among the machine learning groups for their applications in finance and various other fields. Whether it is the asset management, market creation, options trading, or risk management, algorithms of machine learning are being used [1]. Regardless of doubt about past executions, we should now concede that machine learning is altering the financial industry. Almost every segment in the industry has massively invested in machine learning and mostly believes that this is only the start. Indeed, even regulators are intently watching this turn of events and its effect on the sector. Until now the greatest pointer is the improvement of the financial market employing. These days, quant finance applicants should pass certification in machine learning, or at least know about this innovation and have experience in the Python programming language [2].

The stochastic process of general continuous functions is called the Gaussian process (GP), which in other terms can be called the Gaussian random vector generalization. GP replaces the conventional covariance matrix with a kernel function and advantages from the might of the kernel methods. The conditional distribution computation is the central part of this procedure. In the Bayesian orientation structure, this relates to the computation of the latter distribution from the former distribution. It is not difficult to characterize Gaussian process regression, as linear regression is the answer for the conditional expectation issue when the random variable is Gaussian. This is a potent semiparametric model in machine learning. The model has been effectively utilized in the areas of reinforcement research training, multi-task robotics, and geostatistics.

BO is utilized to crack the black-box optimization issues that are not known expressly and are costly to assess. In the event that the gradient vector of the objective function cannot be accessed, the conventional quasi Newton technique or gradient descent methods cannot be utilized. Using an unidentified function as a substitute to a random Gaussian process, BO replaces the unsolvable problems with a sequence of simple optimization problems [2,3]. For this situation, the GPs appear to be an apparatus in BO.

By and large, the financial model relies upon some external parameters that should be modified accordingly prior to beginning of the model. BO basically includes assessing these external parameters called hyperparameters. Instances of these hyperparameters are the periods of moving average assessor, investor risk appetite, and the return on assets (ROA) covariance matrix window.

9.2 Gaussian Processes

In this part, we will break down the fundamental ideas and techniques in utilizing GP in machine learning. GPs are used to identify and choose patterns and to extrapolate and interpolate the problems that are crucial to the decision of kernel functions in classification and regression. GPs have the characteristic of algorithmic learning under a controlled or supervised environment apart from estimating the test sample's confidence range and variance. The function of GP is handy for global optimization and assists to improve the objective function as the test sample can be assessed dependent on the model's confidence.

9.2.1 Gaussian Process Definition

Let X be the set* in R^d. The Gaussian process is the set of $\{f(x), x \in X\}$, so that each $n \in N$ and $x_1...x_n \in X$, the random vector $(f(x_1)....f(x_n))$ has a common multivariate Gaussian distribution, so we can use its mean function to characterize GP in the following way:

$$m(x) = E\left[f(x)\right]$$

and its covariance function:

$$K\left(x,x'\right) = \text{cov}\left(f(x),\ f\left(x'\right)\right)$$

$$= E\left[\left(f(x) - m(x)\right)\left(f(x') - m(x')\right)\right]$$

The Kernel function, $K(x, x')$, which is the covariance function, is vital for GP analysis. The most popular kernel function in machine learning, the exponential kernel**, is given as below [2]:

$$K_{SE}(x, x') = \exp\left(-(1/2)\|x - x'\|_2^2\right)$$

for $x \in R^d$ and $x' \in R^d$

 * *The input data set X can be multivariate (multidimensional)variables (linear regression modeling), time variables (time series forecasting), etc.*
 ** *It is also called RBF (radial basis function) kernel or Gaussian kernel.*

9.2.2 Gaussian Process Regression

For the specified training set of features $\{(x_i, y_i)\}_{i=1}$, the goal of GP regression (GPR) is to find $f(x^*)$ for a quantity of novel inputs $x^* \in R^{n^*xd}$. For the same, we use the Bayesian reference system to calculate the posterior distribution of GP when $x = (x_1 \dots x_n) \in R^{nxd}$.

9.2.3 Covariance Functions

The GP can be viewed as the probability distribution of the function. Therefore, the GP covariance function establishes the function $f(x)$ properties.

For example, $K(x, x') = \min(x, x')$ is the Brownian motion's covariance function, so the GP samples are nowhere differentiable in this kernel function [2].

The regularity, periodicity, and monotonicity of the sample can be restricted by selecting the suitable kernel. In addition, procedures in the kernel can pull out additional intricate patterns and structures from the data. The kernel can be added, multiplied, and subtracted to create another effective covariance function.

9.2.4 Hyperparameter Selection

The covariance function has hyperparameters, such as length scales $\Sigma = \text{diag}(\ell_1 \dots \ell_d)$ in the squared exponential kernel, power α in the rational quadratic kernel, and so on. Every one of these parameters affects how the GP model fits the observed data. It is fixed beforehand, and we can evaluate it.

For this model, we determine the model parameters by θ. The customary method of picking parameters is probability function $L(\theta) = p(y|\theta)$ maximization. The main proposal is the maximization of the sample data y feasibility. In the regression of the GPR, $\theta = (\theta_K, \sigma_\varepsilon)$ is composed of parameters θ_K of the kernel function as well as the standard deviation σ_ε of the noise [3,4].

$$\text{If } z = f(x) \text{ is } \quad GP$$

$$\text{then} \quad p(y|\theta) = \int p(y|\theta, z) p(z|\theta) dz$$

Note: The Bayesian method determines the previous distribution of hyperparameters θ and limits the posterior distribution of GP over θ. However, this does not apply to analytical processing. Also note that if the noise is not Gaussian, the posterior distribution GP is difficult to parse. In both cases, Monte Carlo methods must be used, such as the Hamilton method or the hybrid Monte Carlo method.

9.2.5 Classification

Gaussian regression can also be additionally applied to the classification problems as described below:

For the class index, if a discrete variable is the output. For example, we may want to forecast the price trend of a financial instrument or asset: a confirmatory (positive) return is 1 and other returns are 0.

If we take the example of binary classification, in order to use GP to model the two categories, the sigmoid function $g(x)$ is usually used in front of the data, for instance, the logistic function $logit(x)=(1 +e^{-x})^{-1}$. The output y is

$$\Pr\{y = 1\} = g\big(f(x)\big)$$

$f(x)$ here is the GP on X. The estimated distribution of the new input $x*$ can be marginalized over the latent GP values

$$p\big(z^* \,|\, y, z\big) = \int p\big(z^* \,|\, z\big) p\big(z | y\big) dz$$

where z and z* correspondingly refer to the random variables $f(x)$ and $f(x*)$. We construe that as:

$$\Pr\{y^* = 1 / y\} = \int g\big(z^*\big) p\big(z^* \,|\, z\big) p\big(z | y\big) \ dz^* \, dz - (1)$$

Using the Bayes rule, the posterior distribution $p\,(z|y)$ can be given as:

$$p\big(z | y\big) = \big(p\big(y | z\big) p(z)\big) \big/ p(y)$$

The $p(z * | z)$ above is the customary posterior GP distribution; however, the posterior distribution $p(z|y)$ is not effortless to calculate, so the approximate value is utilized to calculate the integral (1) above. For this purpose, we use two popular methods: Laplace approximation and expectation propagation.

The first method is to use the two-order Taylor expansion around the posterior method around its maximum value; the second method is to deal with uncertainties. The empirical distribution was the observed number to minimize the Kullback–Leibler divergence [4,5].

9.3 Bayesian Optimization

BO is a black-box optimization technique in which few parameters are known about the target function $f(x)$. Generally, BO is very helpful if the evaluation of the function is expensive, when the analytical expressions are not available, or gradient vectors are not available. This is suitable for many complex machine learning tasks that require optimization of hyperparameters [2,3].

For example, it is difficult to calculate an estimate of the structure of a deep neural network if we consider a specific hyperparameters set (as the training model alone may consume a lot of time), and it is impracticable to calculate a gradient vector for each hyperparameter.

9.3.1 General Principles

Bayesian optimization includes two sections:

1. Probability substitute
2. Acquisition function (AF) (or utility function)

Initially, a prior probability model is built for the objective function $f(x)$, and the samples thereafter in $f(x)$ are utilized to obtain the posterior probability distribution (PPD) by updating the probability distribution. This assessment of the objective function is known as a substitute model.

The GP is a famous substitute model for BO on the grounds that the posterior GP is as yet a multivariate normal distribution. At that point, we utilize the utility function to choose new points dependent on this PPD to assess the objective function in the subsequent step. This utility function is known as the AF, we believe it to be the most likely best among the exploitation and exploration options [2,3].

Exploitation refers to samples whose alternative models can predict higher target returns, and exploration can predict samples with higher uncertainties. Subsequently, the overall thought of the BO incorporates the accompanying procedure:

1. Put GP before the objective function $f(x)$.
2. Update the PPD GP on $f(x)$ with every available sample.
3. Use the AF to choose where to take the following measure.
4. Considering the above measure, update the PPD of GP.
5. Repeat stages 2–4 until the approximate maximum value of the objective function $f(x)$ is obtained (or halt the process after a given number of procedure cycles).

9.4 Financial Applications

We utilize GP to replicate and anticipate the yield curve, and utilize BO to build up the online trend-following strategy.

9.4.1 Yield Curve Modeling

To represent the capability of the GP method in finance, let us have a glance first at the US yield curve fitting. The Nelson model is one of the most commonly used models for parameter estimates and significant changes over time [2,4]. GP can be regarded as a non-parametric Bayesian alternative (Figures 9.1 and 9.2).

We can test the GP-based method to predict the movement of the U-shaped curve. This is an exemplary macroeconomic factor that can be utilized as a sign to anticipate the returns on stocks and bonds, so it has practical significance for quantitative asset management. We can use the GP-ARX model. The ARX model assumes that there is a nonlinear relationship between Y_t time series and its prior value in addition to a quantity of external factors X_t [2]

$$Y_t = f\left(Y_{t-1}, Y_{t-2}, \ldots, X_{t-1}, X_{t-2}, \ldots\right) + \varepsilon_t$$

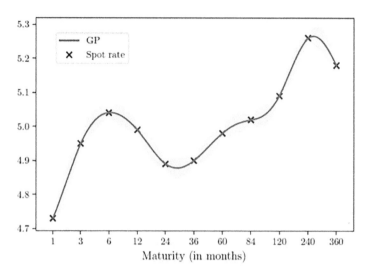

FIGURE 9.1
GP Fitting Yield Curve of spot prices (June 2007).

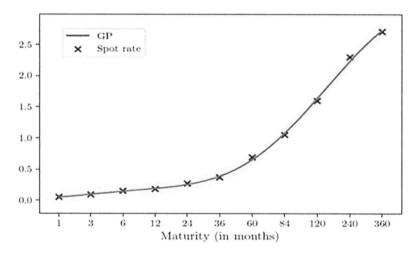

FIGURE 9.2
GP Fitting Yield Curve of spot prices (June 2012).

where $\varepsilon_t \sim N(0, \sigma^2_e)$ is the white noise process.

The primary thought of GP-ARX is to substitute GP for the function $f \sim GP(0, K)$. After selecting the kernel K, the training set only contains the observation values of X_t and Y_t. As depicted in the previous part, the maximum likelihood method is used to perform hyperparameter inference [2,5].

Note: The choice of kernel functions and features is crucial. When the kernel is linear, the ARX model is in fact the same as the GP-ARX model. In the GP-ARX model, we utilize the addition of linear and exponential kernels and just utilize the simple time series of functions. The GP method has the ability to select kernels and can catch distinctive length scales and styles. By finding the right combination of kernels, the GP method can be used as a good interpolator for various financial applications (Figures 9.3 and 9.4).

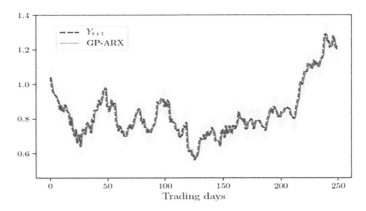

FIGURE 9.3
Two-year spot price forecast.

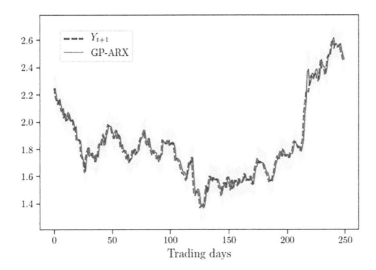

FIGURE 9.4
Ten-year spot price forecast.

9.4.2 Portfolio Optimization

Now, we consider applying BO to portfolio optimization (PO) with regards to quantitative asset management. We initially portray the asset allocation problem as a trend-following strategy, and disclose how to take care of the issue, and lastly utilize the squared exponential kernel of BO to track down the ideal hyperparameters of the trend-following strategy [2,4].

9.4.2.1 Trend Following Strategy

Considering a set of n assets, we notice their everyday prices and find the best portfolio, i.e., the distribution vector $x \in R^n$ that balances risks, and whether it is possible to predict the expected return vector μ and the covariance matrix Σ to calculate the asset return and the regularized Markowitz optimization problem has the following viewpoints:

$$x^*(\gamma) = \arg\min_x \frac{1}{2} x^T \sum x - \gamma \mu^T x + \lambda \|x - x_0\|_2^2$$

Here, γ is the reciprocal of the risk aversion ratio, λ is the peak regularization parameter, and x_0 is the reference portfolio. Let us consider a plain adaptation of the trend-following strategy [2,4]:

- The anticipated rate of return is calculated by means of a moving average estimator. $P_{i,t}$ is the daily asset price i. Then,

$$\mu_{i,t} = \frac{P_{i,t}}{P_{i,t} - \ell(\mu)} - 1$$

  where $\ell(\mu)$ is the length of the estimated window of the MA estimator.
- Use empirical estimation to estimate the covariance matrix, and its window length is denoted as $\ell(\Sigma)$.
- Rebalance the portfolio x_t at a fixed t date (for instance, monthly or weekly). The best possible portfolio at a rebalancing date t is defined using the allocation problem as

$$x_t(\lambda) = \arg\min_x - \mu_t^T x + \lambda \|x - x_{t-1}\|_2^2$$

$$\text{s.t.} \quad \sigma_t(x) \leq \bar{\sigma}$$

Where μ_t is the expected vector of anticipated returns at time t,

$$\sigma_t(x) = \sqrt{x^T \sum_t x}$$

is the estimated portfolio volatility at time t, and σ is the target volatility at which the strategy follows the trend [2,5].

To resolve the above issue, we employ the Alternating Direction Method of Multipliers (ADMM) algorithm to rewrite the problem as follows

$$x_t = \arg\min_x - \mu_t^T x + \lambda \|x - x_{t-1}\|_2^2 + |1_\Omega(z)$$

$$\text{s.t.} \quad x - z = 0$$

where $\Omega = \left\{ z \in \mathbb{R}^n : \left\| z^T \sum_t z \right\|_2^2 \leq \bar{\sigma}^2 \right\}$. However, we improve the ADMM algorithm by introducing the Cholesky trick:

$$x_t = \arg\min_x - \mu_t^T x + \lambda \|x - x_{t-1}\|_2^2 + 1_\Omega(z)$$

$$\text{s.t.} \quad -L_t x + z = 0$$

where $\Omega = \left\{ z \in \mathbb{R}^n : \|z\|_2^2 \leq \bar{\sigma}^2 \right\}$ and L_t is the upper Cholesky decomposition matrix of Σ_t. It follows that $z = L_t x$ and

$$\|z\|_2^2 = z^\mathsf{T} z$$

$$= x^\mathsf{T} L_t^\mathsf{T} L_t x$$

$$= x^\mathsf{T} \sum_t x$$

$$= \sigma_t^2(x)$$

Concerning the formulae of Bourgeron Richard and Roncalli, Cholesky's route aids to pick up the pace of the convergence speed of the ADMM algorithm, which means that the ADMM algorithm transforms as [2,3]:

$$x^{(k+1)} = \arg\min - \mu_t^\mathsf{T} x + \lambda \|x - x_{t-1}\|_2^2 + \frac{\varphi}{2} \left\| -L_t x + z^{(k)} + u^{(k)} \right\|_2^2$$

$$z^{(k+1)} = \arg\min 1_\Omega(z) + \frac{\varphi}{2} \left\| -L_t x^{(k+1)} + z + u^{(k)} \right\|_2^2$$

$$u^{(k+1)} = u^{(k)} - L_t x^{(k+1)} + z^{(k+1)}$$

It should be noted that z-step relates to an uncomplicated projection on the Euclidean sphere centered on the center 0 of the vector $L_t x^{(k+1)} - u^k$ and the radius σ, and the calculation of the near-end operator is not difficult. Step x relates to a linear system. $f(k)(x)$ is as below:

$$f^{(k)}(x) = -\mu_t^\mathsf{T} x + \lambda \|x - x_{t-1}\|_2^2 + \frac{\varphi}{2} \left\| -L_t x + z^{(k)} + u^{(k)} \right\|_2^2$$

We deduce that

$$\nabla f^{(k)}(x) = -\mu_t + \lambda x - \lambda_{x_{t-1}} + \varphi L_t^\mathsf{T} L_t x - \varphi L_t^\mathsf{T} \left(z^{(k)} + u^{(k)} \right)$$

$$= -\mu_t + \lambda x - \lambda_{x_{t-1}} + \varphi \sum_t x - \varphi L_t^\mathsf{T} \left(z^{(k)} + u^{(k)} \right)$$

Finally, we get the following solution:

$$x^{(k+1)} = \left(\varphi \Sigma_t + \lambda I_n \right)^{-1} \left(\mu_t + \lambda x_{t-1} + \varphi L_t^\mathsf{T} \left(z^{(k)} + u^{(k)} \right) \right)$$

9.4.2.2 Hyperparameter Assessment of the Trend-Following Strategy

The trend-following strategy relies upon three hyperparameters:

  i. Parameter λ – the turnover is controlled by parameter λ among two rebalancing dates.
  ii. Window length $\ell(\mu)$ – the trend assessment is controlled by this parameter.
  iii. The horizon time $\ell(\Sigma)$ – risky assets are measured by this parameter.

This strategy is implemented in consideration of fixed hyperparameters [2]. By its very nature, their choice significantly affects the strategy formulation.

For example, a smaller $\ell(\mu)$ value signifies a small time momentum, while a larger $\ell(\mu)$ value shows a further stable trend. The window length hyperparameter is the way to recognize Commodity Trading Advisors (CTAs) of long and short terms.

Indeed, the problem of correct asset allocation is defined by

$$x_t\left(\lambda_t, \ell_t(\mu), \ell_t(\Sigma)\right) = \arg\min_x - \mu_t^T x + \lambda_t \|x - x_{t-1}\|_2^2$$

$$s.t. \sigma_t(x) \leq \bar{\sigma}$$

This implies that the hyperparameters are not set and should be evaluated on every rebalance date. In the prior framework, the assessment is to find x_t as λ, $\ell(\mu)$, and $\ell(\Sigma)$ are constant. In this framework, the assessment likewise incorporates tracking down the best portfolio x_t just as the ideal parameters λ, $\ell(\mu)$, and $\ell(\Sigma)$. This should be possible utilizing the BO framework [2,3].

Parameters $\ell(\mu)$ and $\ell(\Sigma)$ are discrete in nature, and are typically represented in months, e.g. $\ell(\mu) \in \{3,6,12,24\}$ and $\ell(\Sigma) \in \{3,6,12\}$. In BO, discrete, whole number, or categorized parameters are difficult to oversee, in light of the fact that GPs cannot replace black-box functions and therefore cannot be adapted. Since there is no standard method yet, we use a simple method that uses continuous variables in the BO stage and combines the hyperparameters $\ell(\mu)$ and $\ell(\Sigma)$ with the closest whole number to calculate the objective function deduced from the above equation [2,5].

The selection of the objective function is an important step to achieve BO. In classic machine learning problems, the objective function can be a prudently validated estimate of hyperparameters to lessen the possibility of overfitting. The strategy is not very clear, and it is easy to overfit. The Sharpe ratio of this strategy is a simple and an obvious function. At each rebalance date, we run BO to find the hyperparameters that maximize the Sharpe ratio over a period of time [2,4].

A more powerful objective function is to minimize the variable Sharpe ratio to reduce the distortion of overfitting. For example, for a back test with set hyperparameters and period [T-0.5;T], we can calculate the Sharpe ratio $SR_T(\lambda, \ell(\mu), \ell(\Sigma))$ for 6 months, and then use BO to solve

$$\left\{\lambda_t, \ell_t(\mu), \ell_t(\Sigma)\right\} = \arg\max\left\{\min_{\tau \in [t-2,t]} SR_T(\lambda, \ell(\mu), \ell(\Sigma))\right\}$$

However, in order to benefit from the parameter space using BO, we utilize another method. As volatility is as of now constrained by the limitations of PO and regularization, we use a 2-year time frame of the objective function to calculate the strategy [2].

$$\left\{\lambda_t, \ell_t(\mu), \ell_t(\Sigma)\right\} = \arg\max \hat{\mu}_t\left(\lambda, \ell(\mu), \ell(\Sigma)\right)$$

where, $\hat{\mu}_t\left(\lambda, \ell(\mu), \ell(\Sigma)\right)$ is the performance of the back-tested period of $[t-2;t]$.

After tracing all the samples tested in the BO process, they are classified as per their objective function. Then, three best fitted hyperparameter sets are chosen to calculate three distinctive optimal weights. These weights are smoothened to form the concluding investment portfolio. The above process greatly lessens the inclination to overfitting [2,5] (Figure 9.5).

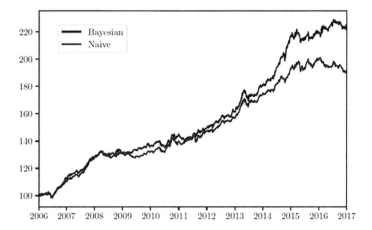

FIGURE 9.5
Cumulative performance of the trend-following strategies.

9.5 Summary

GP is an incredible instrument for optimizing yield curves. Along these lines, GP can be used as a semiparametric option to the well-known parametric methods (such as the Nelson–Siegel model) for the development of online trend-following strategies. Nonetheless, the outcomes additionally show that GPs follow long-established econometric methods to predict interest rates; however, the effect is not satisfactory. The instance of the trend-following strategy is highly intriguing in light of the fact that when we choose ex ante hyperparameters, it brings classic financial problems. So far, the method to test different combinations of hyperparameters and selecting the best combination to avoid sampling bias inherent in back-test protocols is the best known one. We also studied how to execute BO to estimate the window length of the trend vector and covariance matrix.

References

1. Hemachandran K., Rodriguez R.V., Toshniwal R., Junaid M., & Shaw L. (2022) Performance analysis of different classification algorithms for bank loan sectors. In: Raj J.S., Palanisamy R., Perikos I., Shi Y. (eds) *Intelligent Sustainable Systems. Lecture Notes in Networks and Systems*, vol. 213. Springer, Singapore. https://doi.org/10.1007/978-981-16-2422-3_16
2. https://www.researchgate.net/publication/332288851_Financial_Applications_of_Gaussian_Processes_and_Bayesian_Optimization
3. Goodell, J. W., Kumar, S., Lim, W. M., & Pattnaik, D. (2021). Artificial intelligence and machine learning in finance: Identifying foundations, themes, and research clusters from bibliometric analysis. *Journal of Behavioral and Experimental Finance*, 32, 100577. https://doi.org/10.1016/j.jbef.2021.100577

4. Pattnaik, D., Hassan, M. K., Kumar, S., & Paul, J. (2020). Trade credit research before and after the global financial crisis of 2008 – A bibliometric overview. *Research in International Business and Finance*, 54, 101287. https://doi.org/10.1016/j.ribaf.2020.101287

5. Pattnaik, D., Kumar, S., & Vashishtha, A. (2020). Research on trade credit – A systematic review and bibliometric analysis. *Qualitative Research in Financial Markets*, 12(4), 367–390. https://doi.org/10.1108/QRFM-09-2019–0103.

10

Bayesian Network Inference on Diabetes Risk Prediction Data

M. Ö. Cingiz

Bursa Technical University

CONTENTS

10.1 Introduction

Diabetes mellitus is a systematic disorder that causes problems in the eyes, kidneys, nervous system, heart, and vessels over a long period of time [1]. It affected more than 500 million people all over the world in 2020 [2]. Diabetes mellitus has three different types: Type 1 diabetes, Type 2 diabetes, and gestational diabetes. Type 2 diabetes is the most common diabetes type, with early symptoms that are signs of prediabetes. This long-term asymptomatic phase leads to Type 2 diabetes unless prevention steps are taken in the prediabetes period [1]. We work on the early-stage diabetes risk prediction dataset. The dataset consists of signs and symptoms of diagnosed diabetic patients and same symptoms on healthy people. The dataset contains age, gender, polyuria, polydipsia, sudden weight loss, weakness, polyphagia, genital thrush, visual blurring, itching, irritability, delayed healing, partial paresis, muscle stiffness, alopecia, obesity, and class attributes. Class indicates whether the patient is diabetic or healthy. The main subjective of the study is to infer different Bayesian networks (BNs) of diabetes risk prediction dataset to display cause-effect relationship between attributes. There are limited studies in the literature that focus on discovering relations between diabetes symptoms. We intend to derive cause-effect relations between diabetes symptoms and diabetes status via BN structures.

BNs are directed acyclic graphs (DAG) that consist of a set of nodes representing random variables (attributes), are trained probabilistic graphical models from data. BNs are utilized to derive cause- effect relations between variables in many topics such as bioinformatics, health informatics, software engineering, environmental modelling, and social sciences [3–8]. These studies investigate regulator–regulated dependence between variables. Transcription factors and miRNAs regulate gene protein expression level as regulators. Understanding the relation between regulators and gene expression level

DOI: 10.1201/9781003164265-10

enable researchers to develop disease-related drugs as the target for bioinformatics [4,5]. Environmental events also affect each other. Guo [6] found urbanization and grazing have an impact on removal habitat, which causes a decrease in biodiversity. Similar relations are obtained on bug prediction datasets that consist of software metrics. Software metrics can influence other software metrics that cause a different number of software bugs [7]. The cause-effect relations are also inferred in social sciences to understand human behavior in terms of different events [8]. BNs are applicable for many different working areas to discover cause-effect relations.

The second aim of the study is to compare the performance of constraint-based and score-based BN inference algorithms. Score-based algorithms investigate to infer BNs via adding, removing, or reversing arcs between nodes in DAG space due to penalized log-likelihood (LL) functions as score functions. Greedy climbing hill algorithm (GCHA), K2, simulated annealing, and tabu search algorithms are popular search and score algorithms. Other BN inference approaches are constraint-based algorithms such as grow shrink (GS), iterative associate Markov blanket (IAMB), and hybrids of these techniques. These algorithms utilize statistical independence tests (Pearson's chi-squared or Wilk test) to determine conditional independence between random variables to build BN structure. We applied GCHA as a score-based algorithm, IAMB and GS algorithms as constraint-based algorithms to build BNs on early-stage diabetes risk prediction dataset. We intend to plot a cause-effect relationship between attributes using three BN inference algorithms. In the literature there are few studies that compare score-based and constraint-based algorithms on the same datasets. Most of these studies found that constraint-based algorithms can model small datasets more accurately [9–11]. In our chapter, we present methods, results, and discussion in Sections 10.2, 10.3, and 10.4, respectively.

10.2 Methods

BNs are graphical models which consist of nodes and arcs between nodes. This structure forms a DAG, which contains no loop. Nodes represents random variables which have probability values, $P(X)$. Arcs between nodes in BNs are directed and they give information for causality between variables. A directed edge from X variable to Y variable depicts that Y variable is influenced by X variable. The causality between X and Y variables is determined by conditional probabilities, $P(Y|X)$. If the probability value of a variable changes when another variable is known, this variable is called the child variable and the known, given, variable is called the parent variable in BNs. BN techniques learn causal relationships of variables on dataset. BN techniques derive conditional probabilities and apply some constraints to build BNs. We applied constraint-based (conditional independence or Markov blanket learning) and score-based BN inference algorithms in the study. Cause-effect relations of variables are represented as parent–descendant in BNs.

10.2.1 Score-Based BN Inference Algorithms

GCHA is a score-based BN inference algorithm [12,13]. The scoring function reflects how well BNs fit the observed dataset. Random variables can be represented as $V=\{X_1, X_2... X_N\}$, and the conditional probability of X_i, when parent event is given, can be written as

$P(X_i \mid P(X_i))$. The P function symbolizes parent of the variable. The joint distribution of all random variables of the dataset, whose size is N, can be described as in Equation 10.1:

$$P(V) = \prod_{i=1}^{N} P\left(X_i \mid P(X_i)\right) \tag{10.1}$$

Score-based algorithms intend to find an optimal scoring structure for the dataset. The calculation of BN structure score is given in Equation 10.2 by summing conditional probabilities of each random variable of the dataset D:

$$\text{Score (BNS} \mid D) = \sum_{i=1}^{N} \text{Score}\left(X_i \mid P(X_i)\right) \tag{10.2}$$

Score-based algorithms intend to find the highest BN score value, which is given in Equation 10.2. Many scoring functions utilize log-likelihood (LL) functions that are summed with various penalizing terms to avoid overfitting and network complexity. However, maximizing the LL function means minimizing the negative LL function [14].

$$\text{LL}(D \mid \text{BNS}) = \sum_{i=1}^{N} \sum_{j=1}^{n} \log P\left(D_{ij} \mid P_{ij}\right) \tag{10.3}$$

In Equation 10.3, "N" is the number of variables in the dataset and "n" is the number of states for variables. Scoring-based algorithms try to maximize LL values. In other words, we try to minimize negative expected LL function scores. Bnlearn R package can present various model selection metrics that are used to compare the performance results of BN inference algorithms.

We apply the GCHA as a score-based algorithm to infer BN on diabetic dataset. GCHA applies add, delete, and update edge operators to discover the optimal BN structure. Add edge operator adds a directed edge between two nodes. Delete edge operator removes directed edge between two nodes. Update edge operator changes the directions of arcs between two nodes. GCHA cannot guarantee the optimal structure and it may get stuck in a local optimal structure. GCHA steps can be described as below:

1. Determine initial Bayesian Network structure and calculate the score of BN due to LL function (as score function).

2. Assign first BN structure as the optimal BN structure in the first iteration.

3. Apply one of these three operators: add, delete, or update. These operators change the structure of BN.

4. Calculate goodness of fit value of the new BN using LL function. If the score of the new BN is higher than the optimal BN structure, then assign the latest BN as the optimal BN structure.

5. Go to Step 3 until the maximum iteration number is obtained for a stable BN structure.

10.2.2 Constraint-Based Algorithms (Markov Blanket Learning Algorithms)

Markov blanket learning (MBL) algorithms utilize joint probabilities of random variables like GCHA. MBL algorithms assume that each variable is independent from its non-descendants when information of its parent is given. This assumption reduces joint probability distribution over many variables. It prunes redundant directed edges between nodes in BN. Pruning leads to obtaining a simple and robust graphical representation of the variables.

MBL algorithms search three basic independence structures to specify independence between variables.

- Indirect cause: If a parent node X is given, Y (parent of X) and Z (child of X) are independent, $Y \perp\!\!\!\perp Z \mid X$ is given.
- Common effect: Y and Z nodes are parents of node X. Y and Z are independent when parent X is not given. If X node is known then Y and Z become dependent. $Y \perp\!\!\!\perp Z \mid X$ is not given.
- Common cause: If parent node X is given, its children Y and Z become independent. $Y \perp\!\!\!\perp Z \mid X$ is given.

MBL algorithms apply different statistical independence tests such as Pearson's chi-squared test on triplet probabilities, $P(X,Y \mid Z)$. If independence score of triplets, such as $P(X,Y \mid Z)$, is lower than threshold value then X and Y variables are determined as conditional independents.

We used GS and IAMB algorithms as MBL algorithms. IAMB is a variation of GS algorithm. GS algorithm consists of two basic steps: growing and shrinking phases, in that order.

Growing phase steps:

1. For every node, for each iteration take this variable as target variable.
2. Measure independence value of target variable and other variables via independence tests such as Pearson's chi-squared or Wilk test.
3. If any variable is independent from the target variable, add this variable to Markov blanket set of the target (MBS_{target}).
4. Stop the growing phase after applying the above steps for all variables.

Shrinking phase steps:

1. For every target variable and its MBS_{target} set:
2. Remove any variable in MBS_{target}, calculate independence score of selected variable and target variable by evaluating variables in MBS_{target}. If independence score is higher than threshold, then GS removes the variable, which is added to MBS_{target} in the growing phase previously.
3. Apply Growing phase Step 3 for each variable in MBS_{target}.
4. Stop the shrinking phase after applying Step 2 and 3 for all variables.

IAMB is a variant of GS algorithm, and the difference only occurs in the growing phase. IAMB selects the most dependent variable from the target variables rather than making a random selection. This selection criteria avoids missing important variables for independence check. However, GS and IAMB algorithms are popular for BN inference, they have adding and removing dependency problems in high-dimensional datasets.

10.3 Results

We intend to infer cause-effect relations between variables on early-stage diabetes risk prediction dataset. The dataset consists of 17 attributes and 520 instances and it can be accessed via UCI Repository [15]. The dataset consists of signs and symptoms of newly diagnosed diabetic patients or prediabetes. It contains age, gender, polyuria, polydipsia, sudden weight loss, weakness, polyphagia, genital thrush, visual blurring, itching, irritability, delayed healing, partial paresis, muscle stiffness, alopecia, obesity, and class attributes. Class indicates whether the patient is diabetic or healthy.

We applied GCHA, GS ,and IAMB to infer cause-effect relation between 17 variables. Figure 10.1 presents GCHA-based DAG. We used bnlearn *R* package [16] for three BN inference algorithms and Cytoscape [17] for DAG visualization.

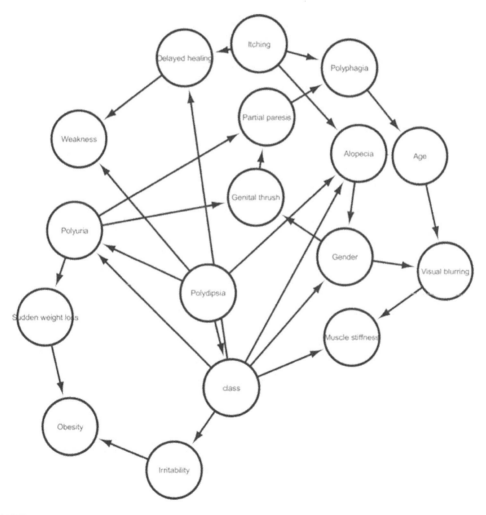

FIGURE 10.1
GCHA-based DAG.

GCHA-based DAG consists of 28 directed edges between all variables and its average neighborhood size is 4.35. Polydipsia has direct effect on the class label as well as polyuria, weakness, and alopecia. Diabetes may cause possibly many variables in dataset. Polydipsia influences polyuria, and age also affects visual blurring. Diabetes directly effects delayed healing, alopecia, muscle stiffness, polyuria, and irritability. Diabetes also has indirect relations such as weakness, visual blurring, genital thrush, sudden weight loss, partial paresis, obesity via delayed healing, gender, polyuria, and irritability. However, if delayed healing, gender, polyuria, and irritability are given as evidence variables, indirect relations between variables become independent from each other due to d-separation.

Figure 10.2 displays GS-based DAG. It has eight directed edges between ten variables, its average neighborhood size and Markov blanket size are 0.94 and 1.18, respectively.

Figure 10.2 does not present class variable as a cause-effect variable. However, it displays relations between ten variables. Visual blurring has direct effect on polyphagia and gender. Weakness and delayed healing also has impact on itching. If itching is given as evidence, weakness and delayed healing also have dependency. This structure is valid for relation between alopecia, polydipsia, and polyuria. Polyuria also causes sudden weight loss according to Figure 10.2.

Figure 10.3 visualizes IAMB-based DAG. It has 12 directed edges between all variables, its average neighborhood size and Markov blanket size are 1.41 and 1.76, respectively.

Figure 10.3 displays that diabetes, class, and variable have a direct impact on polyuria and gender. Class has an indirect relation to genital thrush and sudden weight loss when

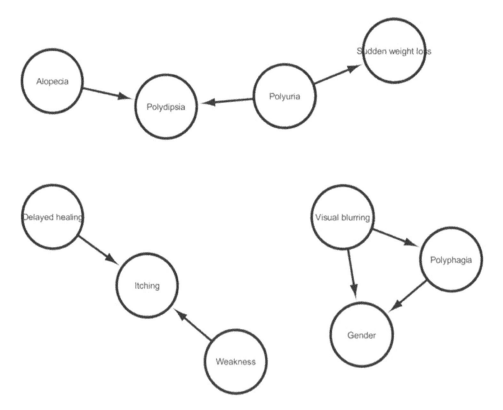

FIGURE 10.2
GS-based DAG.

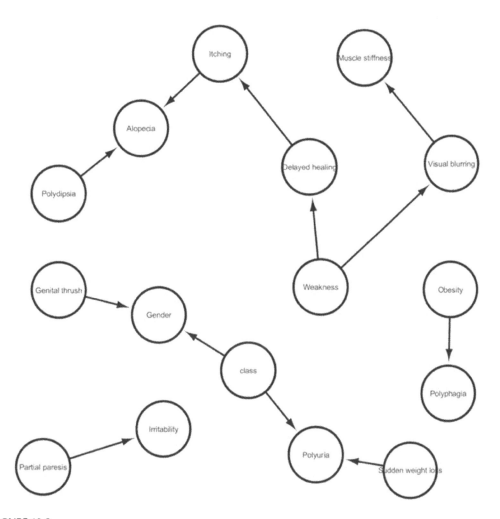

FIGURE 10.3
IAMB-based DAG.

gender and polyuria are given as evidence variables. Weakness causes visual blurring and delayed healing. Polydipsia and itching have an effect on alopecia. Obesity has direct relation to polyphagia and the similar relation is valid from partial paresis to irritability. We can also determine many indirect relations between variables whether evidence variables are given or not for all IAMB-, GS-, and GCHA-based DAGs.

Bnlearn R package also presents different penalized LL score. We used Bayesian information criterion (BIC) (or Schwarz information criterion) to evaluate three different DAGs for model fitness. The BIC formulation is given in Equation 10.4, which penalizes complex models due to its high number of parameters. The BIC score is a negative LL score, which must be minimized.

$$BIC = k\ln(n) - 2\ln(L') \tag{10.4}$$

where L' is the maximized value of the likelihood model, n gives samples, and k is the number of estimated parameters. The lowest BIC value displays how well BN creates a model from the given dataset. BIC values of GCHA-, GS-, and IAMB-based DAGs are −1,991.09, −2,024.09, and −2,072.24, respectively. The lowest value is obtained from IAMB-based DAG due to BIC value. It indicates that IAMB is more successful to infer a BN model on the dataset.

Another penalized LL score is Akaike information criterion (AIC) score, whose calculation is given in Equation 10.5.

$$AIC = 2k(n) - 2\ln(L') \tag{10.5}$$

where L' is the maximized value of likelihood model and k is number of estimated parameters. However, AIC and BIC are utilized for BN model selection; the effect of more parameters is higher for BIC than AIC. The lowest AIC value presents the most appropriate BN model on a given dataset. AIC values of GCHA-, GS-, and IAMB-based DAGs are −1,424.25, −1,754.09, and −1,685.71, respectively. The lowest value is obtained from GS-based DAG due to the AIC value. It indicates that GS is more successful to infer BN model on dataset.

10.4 Discussion

Diabetes mellitus is a complex endemic disease that causes many health problems. It develops gradually for many years and it can be preventable at an early stage. Many patients have also symptoms related to diabetes though they are not aware of their illness. Discovering symptoms of diabetes and understanding the early signs of the disease is important. In our study we used early-stage diabetes risk prediction dataset consisting of 17 attributes and 520 instances. Seventeen variables are assumed signs of diabetes and we want to build the best matching graphical representative model on the dataset via BNs.

BN algorithms can be divided into three groups: score-based algorithms, constraint-based algorithms, and a hybrid of the two approaches. Score-based algorithms try to maximize LL score of BN applying add, delete, and update operators on DAGs. Constraint-based algorithms use conditional independence tests to develop DAGs. We applied GCHA algorithms as score-based algorithms and GS and IAMB as constraint-based algorithms. GCHA-based DAGs infer more cause-effect relations than constraint-based algorithms. The minimum causal relations are derived from GS-based DAG. The three inferred BN structures display the relationship between variables. The impact of diabetes is stated more obviously in GCHA- and IAMB-based DAGs. Diabetes directly causes delayed healing, alopecia, muscle stiffness, polyuria, and irritability, and indirectly effects weakness, visual blurring, genital thrush, sudden weight loss, partial paresis, and obesity according to GCHA-based DAG. IAMB-based DAG also derives relations between class and other variables. However, GS only infers relations between symptom variables. Some inferred relations on DAGs also make sense for human perception. Cause-effect relations between variables such as polyuria–sudden weight loss, age–visual blurring, polyuria–genital thrush, delayed healing–itching, itching–alopecia, obesity–polyphagia can also make sense for human perception.

However, the most dense DAG is derived by GCHA, and model selection scores of constraints-based algorithms are more successful. The minimum BIC score is obtained on IAMB-based DAG and the minimum AIC score is derived on GS-based DAG. There are few studies that compare performance results of constraint-based algorithms and score-based algorithms in the literature. Scutari [9,10] published multiple studies to compare the performance results of constraint-based algorithms and score-based algorithms. Studies revealed that constraint-based algorithms infer more accurate DAGs, minimum BIC score values, than score-based algorithms on small size datasets obtained from simulated and real-world data. Dünder [11] investigated constraint-based algorithm's effect on hybrid BN inference algorithms, and their study states that selection of GS enhances the performance of hybrid algorithm. The results of our study is similar to previous studies in the literature.

We also want to derive cause-effect relations between diabetes and various variables. Another important part of the study is the comparison of different BN inference algorithms on diabetes dataset. The performance and robustness of the study can be increased using many diabetes datasets and hybrid BN inference algorithms in future research. The use of multiple diabetes datasets may infer more robust cause-effect relations. Hybrid algorithms may also increase the performance results of BN-based DAGs.

References

1. Mealey, Brian L., and Gloria L. Ocampo. "Diabetes mellitus and periodontal disease." *Periodontology* 44, 1 (2007): 127–153.
2. Kaiser A.B., Nicole Zhang, and Wouter Van DER PLUIJM. "Global prevalence of type 2 diabetes over the next ten years (2018–2028)." Diabetes 67, Supplement1 (2018):202–LB. doi: 10.2337/db18-202-LB
3. Tamada, Yoshinori, et al. "Estimating gene networks from gene expression data by combining Bayesian network model with promoter element detection." *Bioinformatics* 19, suppl_2 (2003): ii227–ii236.
4. Mhamdi, Hanen, Jérémie Bourdon, Abdelhalim Larhlimi and Mourad Elloumi. Bayesian integrative modeling of genome-scale metabolic and regulatory networks. *Informatics* 7, 1 (2020): 1. Multidisciplinary Digital Publishing Institute.
5. Van den Broeck, Lisa, et al. "Gene regulatory network inference: connecting plant biology and mathematical modeling." *Frontiers in genetics* 11 (2020): 457.
6. Guo, Kai, et al. "A spatial Bayesian-network approach as a decision-making tool for ecological-risk prevention in land ecosystems." *Ecological Modelling* 419(2020): 108929.
7. Okutan, Ahmet, and Olcay Taner Yıldız. "Software defect prediction using Bayesian networks." *Empirical Software Engineering* 19, 1 (2014): 154–181.
8. Jackman, Simon. *Bayesian analysis for the social sciences.* Vol. 846. John Wiley & Sons (2009). ISBN:9780470686638, 0470686634
9. Scutari, Marco, Catharina Elisabeth Graafland, and José Manuel Gutiérrez. "Who learns better bayesian network structures: Constraint-based, score-based or hybrid algorithms?" *International Conference on Probabilistic Graphical Models.* PMLR, 2018.
10. Scutari, Marco, Catharina Elisabeth Graafland, and José Manuel Gutiérrez. "Who learns better Bayesian network structures: Accuracy and speed of structure learning algorithms." *International Journal of Approximate Reasoning* 115 (2019): 235–253.
11. Dünder, Emre, Mehmet Ali Cengiz, and Haydar Koç. "Investigation of the impacts of constraint-based algorithms to the quality of Bayesian network structure in hybrid algorithms for medical studies." *Journal of Advanced Scientific Research* 5, 1 (2014): 08–12.
12. Kask, Kalev, and Rina Dechter. "Stochastic local search for Bayesian network." *AISTATS*, R2(1999): 1–10.

13. Tsamardinos, Ioannis, Laura E. Brown, and Constantin F. Aliferis. "The max-min hill-climbing Bayesian network structure learning algorithm." *Machine Learning* 65, 1 (2006): 31–78.
14. Liu, Zhifa, Brandon Malone, and Changhe Yuan. "Empirical evaluation of scoring functions for Bayesian network model selection." *BMC Bioinformatics* 13, 15 (2012): 1–16.
15. Islam, MM Faniqul, et al. "Likelihood prediction of diabetes at early stage using data mining techniques." *Computer Vision and Machine Intelligence in Medical Image Analysis* (2020): 113–125. doi: 10.1007/978-981-13-8798-2_12
16. Scutari, Marco. "Learning Bayesian networks with the bnlearn R package." *arXiv Preprint arXiv:0908.3817* (2009): 1–22.
17. Saito, Rintaro, et al. "A travel guide to Cytoscape plugins." *Nature Methods* 9, 11 (2012): 1069.

Index

Note: Italic page numbers refer to figures.

Taylor & Francis Group
an **informa** business

Taylor & Francis eBooks

www.taylorfrancis.com

A single destination for eBooks from Taylor & Francis
with increased functionality and an improved user
experience to meet the needs of our customers.

90,000+ eBooks of award-winning academic content in
Humanities, Social Science, Science, Technology, Engineering,
and Medical written by a global network of editors and authors.

TAYLOR & FRANCIS EBOOKS OFFERS:

A streamlined
experience for
our library
customers

A single point
of discovery
for all of our
eBook content

Improved
search and
discovery of
content at both
book and
chapter level

REQUEST A FREE TRIAL
support@taylorfrancis.com

Routledge
Taylor & Francis Group

CRC Press
Taylor & Francis Group

Milton Keynes UK
Ingram Content Group UK Ltd.
UKHW050440111024
449327UK00038B/166